The Cultivation of The Nati
Americ

by George Husmann

INTRODUCTION

It is with a great deal of hesitation I undertake to write a book about Grapes, a subject which has been, and still is, elucidated every day; and about which we have already several works, which no doubt are more learned, more elaborate, than anything I may produce. But the subject is of such vast importance, and the area suitable for grape culture so large, the diversity of soil and climate so great, that I may be pardoned if I still think that I could be of some use to the beginner; it is for them, and not for my brethren of the craft more learned than I am, that I write. If they can learn anything from the plain talk of a practical worker, to help them along in the good work, I am well repaid.

Another object I have in view is to make grape growing as easy as possible; and I may be pardoned if I say that, in my opinion, it is a defect in all books we have on grape culture, that the manner of preparing the soil, training, etc., are on too costly a plan to be followed by men of little means. If we are first to trench and prepare the soil, at a cost of about $300 per acre, and then pay $200 more for trellis, labor, etc., the poor man, he who must work for a living, can not afford to raise grapes. And yet it is from the ranks of these sturdy sons of toil that I would gain my recruits for that peaceful army whose sword is the pruning-hook; it is from their honest, hard-working hands I expect the grandest results. He who has already wealth enough at command can of course afford to raise grapes with bone-dust, ashes, and all the fertilizers. He can walk around and give his orders, making grape culture an elegant pastime for his leisure hours, as well as a source of profit. But, being one of the first class myself, I had to fight my way up through untold difficulties from the lowest round of the ladder; had to gain what knowledge I possess from dear experience, and can therefore sympathize with those who must commence without means. It is my earnest desire to save them some of the losses which I had to suffer, to lighten their toil by a little plain advice. If I can succeed in this, my object is accomplished.

In nearly all our books on grape culture I notice another defect, especially in those published in the East; it is, that they contain a great deal of good advice about grape culture, but very little about wine-making, and the treatment of wine in the cellar. For us here at the West this is an all-important point, and even our Eastern friends, if they continue to plant grapes at the rate they have done for the last few years, will soon glut the market, and will be forced to make them into wine. I shall therefore try to give such simple instructions about wine-making and its management as will enable every one to make a good saleable and drinkable wine, better than nine-tenths of the foreign wines, which are now sold at two to three dollars per bottle. I firmly believe that this continent is destined to be the greatest wine-producing country in the world; and that the time is not far distant when wine, the most wholesome and purest of all stimulating drinks, will be within the reach of the common laborer, and take the place of the noxious and poisonous liquors which are now the curse of so many of our laboring men, and have blighted the happiness of so many homes. Pure light wine I consider the best temperance agent; but as long as bad whisky and brandy continue to be the common drink of its citizens we can not hope to accomplish a thorough reform; for human nature seems to crave and need a stimulant. Let us then try to supply the most innocent and healthy one, the exhilarating juice of the grape.

I have also endeavored throughout to give plain facts, to substantiate with plain figures all I assert; and in no case have I allowed fancy to roam in idle speculations which cannot be demonstrated in practice. I do not pretend that my effort is "the most comprehensive and practical essay on the grape," as some of our friends call their productions, but I can claim for it strict adherence to truth and actual results.

I have not thought it necessary to give the botanical description of the grape-vine, and the process of hybridizing, etc.; this has already been so well and thoroughly done by my friend FULLER, that I can do no better than refer the scientific reader to his book. I am writing more for the practical farmer, and would rather fill what I think a vacancy, than repeat what has been so well said by others.

With these few remarks, which I thought due to the public and myself, I leave it to you, brother-winegrowers, to say whether or not I have accomplished my task. To all and every one who plants a single vine I would extend the hand of good fellowship, for he is a laborer in the great work to cover this glorious land of the free with smiling vineyards, and to make its barren spots flow with noble grape juice, one of the best gifts of an all-bountiful Creator. All hail to you, I greet you from Free Missouri.

GRAPE CULTURE

REMARKS ON ITS HISTORY IN AMERICA, ESPECIALLY AT THE WEST--ITS PROGRESS AND ITS FUTURE.

In an old chronicle, entitled, "The Discovery of America in the Tenth Century," by CHARLES C. PRASTA, published at Stralsund, we find the following legend:

"LEIF, son of ERIC the Red, bought BYARNES' vessel, and manned it with thirty-five men, among whom was also a German, TYRKER by name, who had lived a long time with LEIF'S father, who had become very much attached to him in youth. And they left port at Iceland, in the year of our Lord 1000.

But, when they had been at sea several days, a tremendous storm arose, whose wild fury made the waves swell mountain high, and threatened to destroy the frail vessel. And the storm continued for several days, and increased in fury, so that even the stoutest heart quaked with fear; they believed that their hour had come, and drifted along at the mercy of wind and waves. Only LEIF, who had lately been converted to CHRIST our Lord, stood calmly at the helm and did not fear; but called on Him who had walked the water and quieted the billows, with firm faith, that He also had power to deliver them, if they but trusted in Him. And, behold! while he still spoke to them of the wonderful deeds of the Lord, the clouds cleared away, the storm lulled; and after a few hours the sea, calmed down, and rocked the tired and exhausted men into a deep and calm sleep. And when they awoke, the next morning, they could hardly trust their eyes. A beautiful country lay before them, green hills, covered with beautiful forests--a majestic stream rolled its billows into the ocean; and they cast the anchor, and thanked the Lord, who had delivered them from death.

A delightful country it seemed, full of game, and birds of beautiful plumage; and when they went ashore, they could not resist the temptation to explore it. When they returned, after several hours, TYRKER alone was missing. After waiting some time for his return, LEIF, with twelve of his men, went in search of him. But they had not gone far, when they met him, laden down with grapes. Upon their enquiry, where he had stayed so long, he answered: "I did not go far, when I found the trees all covered with grapes; and as I was born in a country, whose hills are covered with vineyards, it seemed so much like home to me, that I stayed a while and gathered them." They had now a twofold occupation, to cut timber, and gather grapes; with the latter, they loaded the boat. And Leif gave a name to the country, and called it Vinland, or Wineland."

So far the tradition. It is said that coming events cast their shadows before them. If this is so, may we not recognize one of those shadows in the old Norman legend of events which transpired more than eight hundred years ago? Is it not the foreshadowing of the destiny of this great continent, to become, in truth and verity, a Wineland. Truly, the results of to-day would certainly justify us in the assertion, that there is as much, nay more, truth than fiction in it. Let us take a glance at the first commencement of grape culture, and see what has been the progress in this comparatively new branch of horticulture.

From the very first settlement of America, the vine seems to have attracted the attention of the colonists, and it is said that as early as 1564, wine was made from the native grape in Florida. The earliest attempt to establish a vineyard in the British North American Colonies was by the London Company in Virginia, about the year 1620; and by 1630, the prospect seems to have been encouraging enough to warrant the importation of several French vine-dressers, who, it is said, ruined the vines by bad treatment. Wine was also made in Virginia in 1647, and in 1651 premiums were offered for its production. BEVERLY even mentions, that prior to 1722, there were vineyards in that colony, producing seven hundred and fifty gallons per year. In 1664, Colonel RICHARD NICOLL, Governor of New York, granted to PAUL RICHARDS, a privilege of making and selling wine free of all duty, he having been the first to enter upon the cultivation of the vine on a large scale. BEAUCHAMP PLANTAGENET, in his description of the province of New Albion, published in London, in 1648, states "that the English settlers in Uvedale, now Delaware, had vines running on mulberry and sassafras trees; and enumerates four kinds of grapes, namely: Thoulouse Muscat, Sweet Scented, Great Fox, and Thick Grape; the first two, after five months, being boiled and salted and well fined, make a strong red Xeres; the third, a light claret; the fourth, a white grape which creeps on the land, makes a pure, gold colored wine. TENNIS PALE, a Frenchman, out of these four, made eight sorts of excellent wine; and says of the Muscat, after it had been long boiled, that the second draught will intoxicate after four months old; and that here may be gathered and made two hundred tuns in the vintage months, and that the vines with good cultivation will mend." In 1633, WILLIAM PENN attempted to establish a vineyard near Philadelphia, but without success. After some years, however, Mr. TASKER, of Maryland, and Mr. ANTIL, of Shrewsbury, N.J., seem to have succeeded to a certain extent. It seems, however, from an article which Mr. ANTIL wrote of the culture of the grape, and the manufacture of wine, that he cultivated only foreign varieties.

In 1796, the French settlers in Illinois made one hundred and ten hogsheads of strong wine from native grapes. At Harmony, near Pittsburgh, a vineyard of ten acres was planted by FREDERIC RAPP, and his associates from Germany; and they continued to cultivate grapes and silk, after their removal to another Harmony in Indiana.

In 1790, a Swiss colony was founded, and a fund of ten thousand dollars raised in Jessamine county, Kentucky, for the purpose of establishing a vineyard, but failed, as they attempted to plant the foreign vine. In 1801, they removed to a spot, which they called Vevay, in Switzerland County, Indiana, on the Ohio, forty-five miles below Cincinnati. Here they planted native vines, especially the Cape, or Schuylkill Muscadel, and met with better success. But, after about forty years' experience, they seem to have become discouraged, and their vineyards have now almost disappeared.

These were the first crude experiments in American grape culture; and from some cause or another, they seem not to have been encouraging enough to warrant their continuation. But a new impetus was given to this branch of industry, by the introduction of the Catawba, by Major ADLUM, of Georgetown, D.C., who thought, that by so doing, he conferred a greater benefit upon the nation than he would have done, had he paid the national debt. It seems to have been planted first on an extensive scale by NICHOLAS LONGWORTH, near Cincinnati, whom we may justly call one of the founders of American grape culture. He adopted the system of leasing parcels of unimproved land to poor Germans, to plant with vines; for a share, I believe, of one-half of the proceeds. It was his ambition to make the Ohio the Rhine of America, and he has certainly done a good deal to effect it. In 1858, the whole number of acres planted in grapes around Cincinnati, was estimated, by a committee appointed for that purpose, at twelve hundred acres, of which Mr. LONGWORTH owned one hundred and twenty-two and a half acres, under charge of twenty-seven tenants. The annual produce was estimated by the committee at no less than two hundred and forty thousand gallons, worth about as many dollars then. We may safely estimate the number of acres in cultivation there now, at two thousand. Among the principal grape growers there, I will mention Messrs. ROBERT BUCHANAN, author of an excellent work on grape culture, MOTTIER, BOGEN, WERK, REHFUSS, DR. MOSHER, etc.

Well do I remember, when I was a boy, some fourteen years old, how often my father would enter into conversation with vintners from the old country, about the feasibility of grape culture in Missouri. He always contended that grapes should succeed well here, as the woods were full of wild grapes, some of very fair quality, and that this would indicate a soil and climate favorable to the vine. They would ridicule the idea, and assert that labor was too high here, even if the vines would succeed, to make it pay; but they could not shake his faith in the ultimate success of grape culture. Alas! he lived only long enough to see the first dawnings of that glorious future which he had so often anticipated, and none entered with more genuine zeal upon the occupation than he, when an untimely death took him from the labor he loved so well, and did not even allow him to taste the first fruits of the vines he had planted and fostered. Had he been spared until now, his most sanguine hopes would be verified.

I also well remember the first cultivated grape vine which produced fruit in Hermann. It was an Isabella, planted by a Mr. FUGGER, on the corner of Main and Schiller streets, and trained over an arbor. It produced the first crop in 1845, twenty years ago, and so plentifully did it bear, that several persons were encouraged by this apparent success, to plant vines. In 1846, the first wine was made here, and agreeably surprised all who tried it, by its good quality. The Catawba had during that time, been imported from Cincinnati, and the first partial crop from it, in 1848, was so plentiful, that every body, almost, commenced planting vines, and often in very unfavorable localities. This, of course, had a bad influence on so capricious a variety as the Catawba; rot and mildew appeared, and many became discouraged, because they did not realize what they had anticipated. A number of unfavorable seasons brought grape growing almost to a stand still here. Some of our most enterprising grape growers still persevered, and succeeded by careful treatment, in making even the Catawba pay very handsome returns.

It was about this time, that the attention of some of our grape-growers was drawn towards a small, insignificant looking grape, which had been obtained by a Mr. WIEDERSPRECKER from Mr. HEINRICHS, who had brought it from Cincinnati, and, almost at the same time, by Dr. KEHR, who had brought it with him from Virginia. The vine seemed a rough customer, and its fruit very insignificant when compared with the large bunch and berry of the Catawba, but we soon observed that it kept its foliage bright and green when that of the Catawba became sickly and dropped; and also, that no rot or mildew damaged the fruit, when that of the Catawba was nearly destroyed by it. A few tried to propagate it by cuttings, but generally failed to make it grow. They then resorted to grafting and layering, with much better success. After a few years a few bottles of wine were made from it, and found to be very good. But at this time it almost received its death-blow, by a very unfavorable letter from Mr. LONGWORTH, who had been asked his opinion of it, and pronounced it worthless. Of course, with the majority, the fiat of Mr. LONGWORTH, the father of American grape-culture, was conclusive evidence, and they abandoned it. Not all, however; a few persevered, among them Messrs. JACOB ROMMEL, POESCHEL, LANGENDOERFER, GREIN, and myself. We thought Mr. LONGWORTH was human, and might be mistaken; and trusted as much to the evidence of our senses as to his verdict, therefore increased it as fast as we could, and the sequel proved that we were right. After a few years more wine was made from it in larger quantities, found to be much better than the first imperfect samples; and now that despised and condemned grape is the great variety for red wine, equal, if not superior to, the best Burgundy and Port; a wine of which good judges, heavy importers of the best European wines too, will tell you that it has not its equal among all the foreign red wines; which has already saved the lives of thousands of suffering children, men, and women, and therefore one of the greatest blessings an all-merciful God has ever bestowed upon suffering humanity. This despised grape is now the rage, and 500,000 of the plants could have been sold from this place alone the last fall, if they could have been obtained. Need I name it? it is the Norton's Virginia. Truly, "great oaks from little acorns grow!" and I boldly prophecy to-day that the time is not far distant when thousands upon thousands of our hillsides will be covered with its luxuriant foliage, and its purple juice become one of the exports to Europe; provided, always, that we do not grow so fond of it as to drink it all. I think that this is pre-eminently a Missouri grape. Here it seems to have found the soil in which it flourishes best. I have seen it in Ohio, but it does not look there as if it was the same grape. And why should it? They drove it from them and discarded it in its youth; we fostered it, and do you not think, dear reader, there sometimes is gratitude in plants as well as in men? Other States may plant it and succeed with it, too, to a certain extent, but it will cling with the truest devotion to those localities where it was cared for in its youth. Have we not also found, during the late war, that the Germans, the adopted citizens of this great country, clung with a heartier devotion to our noble flag, and shed their blood more freely for it, than thousands upon thousands of native-born Americans? And why? Because here they found protection, equal rights for all, and that freedom which had been the idol of their hearts, and haunted their dreams by night; because they had been oppressed so long they more fully appreciated the blessings of a free government than those who had enjoyed it from their birth. But you may call me fantastical for comparing plants to human beings, and will say, plants have no appreciation of such things. Brother Skeptic, have you, or has any body, divined all the secrets of Nature's workshop? Truly we may say that we have not, and we meet with facts every day which are stranger than fiction.

The Concord had as small a beginning with us. In the winter of 1855 a few eyes of its wood were sent me by Mr. JAS. G. SOULARD, of Galena, Ill. I grafted them upon old Catawba vines, and one of them grew. The next year I distributed some of the scions to our vine-growers, who grafted them also. When my vine commenced to bear I was astonished, after what I had heard of the poor quality of the fruit from the East, to find it so fine, and so luxurious and healthy; and we propagated it as fast as possible. Now, scarcely nine years from the time when I received the first scions, hundreds of acres are being planted with it here, and one-third of an acre of it, planted five years ago, has produced for me, in fruit, wine, layers, cuttings, and plants, the round sum of ten thousand dollars during that time. Its wine, if pressed as soon as the grapes are mashed, is eminently one of those which "maketh glad the heart of man," and is evidently destined to become one of the common drinks of our laboring classes. It is light, agreeable to the palate, has a very enlivening and invigorating effect, and can be grown as cheap as good cider. I am satisfied that an acre will, with good cultivation, produce from 1,000 to 1,500 gallons per year. My vines produced this season at the rate of 2,500 gallons to the acre, but this may be called an extra-large crop. I have cited the history of these two varieties in our neighborhood merely as examples of progress. It would lead too far here, to follow the history of all our leading varieties, though many a goodly story might be told of them. Our friends in the East claim as much for the Delaware and others, with which we have not been able to succeed. And here let me say that the sooner we divest ourselves of the idea that one grape should be the grape for this immense country of ours; the sooner we try to adapt the variety to the locality--not the locality to the variety--the sooner we will succeed. The idea is absurd, and unworthy of a thinking people, that one variety should succeed equally well or ill in such a diversity of soil and climate as we have in this broad land of ours. It is in direct conflict with the laws of vegetable physiology, as well as with common sense and experience. In planting our vineyards we should first go to one already established, which we think has the same soil and location, or nearly so, as the one we are going to plant. Of those varieties which succeed there we should plant the largest number, and plant a limited number also of all those varieties which come recommended by good authority. A few seasons will show which variety suits our soil, and what we ought to plant in preference to all others. Thus the Herbemont, the Cynthiana, Delaware, Taylor, Cunningham, Rulander, Martha, and even the Iona, may all find their proper location, where each will richly reward their cultivator; and certainly they are all too good not to be tried.

Now, let us see what progress the country at large has made in grape-growing during, say, the last ten years. Then, I think I may safely assert, that the vineyards throughout the whole country did not comprise more than three to four thousand acres. Now I think I may safely call them over two millions of acres. Then, our whole list embraced about ten varieties, all told, of which only the Catawba and Isabella were considered worthy of general cultivation; now we count our native varieties by the hundreds, and the Catawba and Isabella will soon number among the things which have been. Public taste has become educated, and they are laid aside in disgust, when such varieties as the Herbemont, Delaware, Clara, Allen's Hybrid, Iona, Adirondac, and others can be had. Then, grape-growing was confined to only a few small settlements; now there is not a State in the Union, from Maine to California, but has its vineyards; and especially our Western States have entered upon a race which shall excel the other in the good work. Our brethren in Illinois bid fair to outdo us, and vineyards spring up as if by magic, even on the prairies. Nay, grape-culture bids fair to extend into Minnesota, a country which was considered too cold for almost anything except oats, pines, wolves, bears, and specimens of daring humanity encased in triple wool. We begin to find out that we have varieties which will stand almost anything if they are only somewhat protected in winter. It was formerly believed that only certain favored locations and soils in each State would produce good grapes--for instance, sunny hillsides along large streams; now we begin to see that we can grow some varieties of grape on almost any soil. One of the most flourishing vineyards I have ever seen is on one of the islands in the Missouri river, where all the varieties planted there--some six or seven--seemed perfectly at home in the rich, sandy mould, where it needs no trenching to loosen the soil. Then, grape-growing, with the varieties then in cultivation, was a problem to be solved; now, with the varieties we have proved, it is a certainty that it is one of the most profitable branches of horticulture, paying thousands of dollars to the acre every year. Then, wine went begging at a dollar a gallon; now it sells as fast as made at from two dollars to six dollars a gallon. Instead of the only wine then considered fit to drink, we number our wine-producing varieties by the dozen, all better than the Catawba; among the most prominent of which I will name--of varieties producing white wine, the Herbemont, Delaware, Cassidy, Taylor, Rulander, Cunningham, and Louisiana; of light-red wines, the Concord; of dark-red wines, the Norton's Virginia, Cynthiana, Arkansas and Clinton; so that every palate can be suited. And California bids fair to outdo us all; for there, I am told, several kinds of wine are made from the same grape, in the same vineyard, and in fabulous quantities. To cite an example of the increase in planting: in 1854 the whole number of vines grown and sold in Hermann did not exceed two thousand. This season two millions of plants have been grown and sold, and not half enough to meet the demand. It is said that the tone of the press is a fair indication of public sentiment. If this is true what does it prove? Take one of our horticultural periodicals, and nine-tenths of the advertisements will be "Grape-vines for sale," in any quantity and at any price, from five dollars to one hundred dollars per 100, raised North, East, South, and West. Turn to the reading matter, and you can hardly turn over a leaf but the subject of grapes stares you in the face, with a quiet impunity, which plainly says, "The nation is affected with grape fever; and while our readers have grape on the brain there is no fear of overdosing." Why, the best proof I can give my readers that grape fever does exist to an alarming degree, is this very book itself. Were not I and they affected with the disease, I should never have presumed to try their patience.

But, fortunately, the remedy is within easy reach. Plant grapes, every one of you who is thus afflicted, until our hillsides are covered with them, and we have made our barren spots blossom as the rose.

Truly, the results we have already obtained, are cheering enough. And yet all this has been principally achieved in the last few years, while the nation was involved in one of the most stupendous struggles the world ever saw, while its very existence was endangered, and thousands upon thousands of her patriotic sons poured out their blood like water, and the husbandman left his home; the vintner his vineyard, to fight the battles of his country. What then shall we become now, when peace has smiled once more upon our beloved country; and the thousands of brave arms, who brandished the sword, sabre, or musket, have come home once more; and their weapons have been turned into ploughshares, and their swords into pruning hooks? When all the strong and willing hands will clear our hillsides, and God's sun shines upon one great and united people; greater and more glorious than ever; because now they are truly free. Truly the future lies before us, rich in glorious promise; and ere long the words and the prophecy contained in the old legend will become sober truth, and America will be, from the Atlantic to the Pacific one smiling and happy Wineland; where each laborer shall sit under his own vine, and none will be too poor to enjoy the purest and most wholesome of all stimulants, good, cheap, native wine. Then drunkenness, now the curse of the nation, will disappear, and peace and good will towards all will rule our actions. And we, brother grape growers? Ours is this great and glorious task; let us work unceasingly, with hand, heart, and mind; truly the object is worthy of our best endeavors. Let those who begin to-day, remember how easy their task with the achievements and experiments of others before them, compared with the labors of those who were the pioneers in the cultivation of the vine.

PROPAGATION OF THE VINE.

I.--FROM SEED.

This would seem to be the most natural mode, were not the grape even more liable to sport than almost any other fruit. It is, however, the only method upon which we can depend for obtaining new and more valuable varieties than we already possess, and to which we are already indebted for all the progress made in varieties, a progress which is, indeed, very encouraging; for who would deny that we are to-day immeasurably in advance of what we were ten years ago. Among the innumerable varieties which spring up every day, and which find ready purchasers, just because they are new, there are certainly some of decided merit. But those who grow seedlings, should bear in mind, that the list of our varieties is already too large; that it would be better if three-fourths of them were stricken off, and that no new variety should be brought before the public, unless it has some decided superiority over any of the varieties we already have, in quality, productiveness and exemption from disease. It is poor encouragement to the grape growing public, to pay from two to five dollars a vine for a new variety, with some high-sounding name, if, after several years of superior cultivation and faithful trial, they find their costly pet inferior to some variety they already possessed, and of which the plants could be obtained at a cost of from ten to fifty cents each.

The grapes from which the seed is to be used, should be fully ripe, and none but well developed, large berries, should be taken. Keep these during the winter, either in the pulp, or in cool, moist sand, so that their vitality may remain unimpaired. The soil upon which your seed-bed is made, should be light, deep and rich, and if it is not so naturally, should be made so with well decomposed leaf-mould. As soon as the weather in spring will permit, dig up the soil to the depth of at least eighteen inches, pulverising it well; then sow the seed in drills, about a foot apart, and about one inch apart in the rows, covering them about three-quarters of an inch deep. It will often be found necessary to shade the young plants when they come up, to prevent the sun from scalding them, but this should not be continued too long, as the plants will become too tender, if protected too long. When the young plants have grown about six inches, they may be supplied with small sticks, to which they will cling readily; the ground should be kept clean and mellow, and a light mulch should be applied, which will keep the soil loose and moist. The young plants should be closely watched, and if any of them show signs of disease, they should at once be pulled up; also those which show a very feeble and delicate growth; for we should only try to grow varieties with good, healthy constitutions. In the Fall, the young plants should be either taken up, and carefully heeled in, or they should be protected by earth, straw, or litter thrown over them. In the Spring, they may be transplanted to their permanent locations; the tops shortened in to six inches, and the roots shortened in to about six inches from the stem. The soil for their reception should be moderately light and rich, and loosened up to the depth of at least eighteen inches.

Make a hole about eight inches deep, then throw in soil so as to raise a small mound in the centre of the hole, about two inches high; on this place the young vine, and carefully spread the roots in all directions; then fill up with well pulverized soil, so that the upper eye or bud is even with the surface of the ground; then press the soil down lightly; place a good stake, of about four feet high, with the plant, and allow but one shoot to grow, which should be neatly tied to the stake as it grows. The vines may be planted in rows six feet apart, and three feet apart in the rows, as many of them will prove worthless, and have to be taken out. Allow all the laterals to grow on the young cane, as this will make it short-jointed and stocky. Cultivate the ground well, stirring it freely with plough, cultivator, hoe, and rake, which generally is the best mulch that can be applied.

With the proper care and attention, our seedlings will generally grow from three to four feet, and make stout, short-jointed wood this second season. Should any of them look particularly promising, fruit may be obtained a year sooner by taking the wood of it, and grafting strong old vines with it. These grafts will generally bear fruit the next season. The method to be followed will be given in another place.

At the end of the second season the vines should be pruned to about three eyes or buds, and the soil hilled up around them so as to cover them up completely. The next spring take off the covering, and when the young shoots appear allow only two to grow. After they have grown about eighteen inches, pinch off the top of the weakest, so as to throw the growth into the strongest shoot, which keep neatly tied to the stake, treating it as the summer before, allowing all the laterals to grow. Cultivate the soil well. At the end of this season's growth the vines should be strong enough to bear the following summer. If they have made from eight to ten feet of stocky growth, the leading cane may be pruned to ten or twelve eyes, and the smaller one to a spur of two eyes. If they will fruit at all, they will show it next summer, when only those promising well should be kept, and the barren and worthless ones discarded.

II.--BY SINGLE EYES.

As this method is mostly followed only by those who propagate the vine for sale in large quantities, and but to a limited extent by the practical vineyardist, I will give only an outline of the most simple manner, and on the cheapest plan. Those wishing further information will do well to consult "The Grape Culturist," by Mr. A. S. FULLER, in which excellent work they will find full instructions.

The principal advantages of this mode of propagation are the following: 1st. The facility with which new and rare kinds can be multiplied, as every well ripened bud almost can be transformed into a plant. 2d. As the plants are started under glass, by bottom heat, it lengthens the season of their growth from one to two months. 3d. Every variety of grape can be propagated by this method with the greatest ease, even those which only grow with the greatest difficulty, or not at all, from cuttings in open ground.

As to the merits or demerits of plants grown under glass from single eyes, to those grown from cuttings or layers in open ground, opinions differ very much, and both have their advocates. For my part, I do not see why a plant grown carefully from a single eye should not be as good as one propagated by any other method; a poor plant is not worth having, whether propagated by this or any other method, and, unfortunately, we have too many of them.

THE PROPAGATING HOUSE.

I will only give a description of a lean-to of the cheapest kind, for which any common hot-bed sash, six feet long, can be used.

Choose for a location the south side of a hill, as, by making the house almost entirely underground, a great deal of building material can be saved. Excavate the ground as for a cellar--say five feet deep on the upper side, seven feet wide, and of any length to suit convenience, and the number of plants you wish to grow. Inside of the excavation set posts or scantlings, the upper row to be seven feet long above the ground, and two feet below the ground; the lower row four and one-half feet above the ground, so that the roof will have about two and one-half feet pitch. Upon these nail the rafters, of two-inch planks. Then take boards, say common inch-plank, and set them up behind the posts, one above the other, to prevent the earth from falling in. This will make all the wall that is needed on both sides. On the ends, boards can be nailed to both sides of the posts, and the intervening space tilled with spent tan or saw-dust. Upon the rafters place the sash on the lower side; the upper side may be covered with boards or shingles, where also the ventilating holes can be left, to be closed with trap-doors. The house is to be divided into two compartments--the furnace-room on one end, about eight feet long, and the propagating house, The furnace is below the ground, say four feet long, the flue to be made of brick, and to extend under the whole length of the bench. To make the flue, lay a row of bricks flat and crosswise; on the ends of these place two others on their edges, and across the top lay a row flat, in the same way as the bottom ones were placed. This gives the flue four inches by eight in the clear. The flue should rise rather abruptly from the furnace, say about a foot; it can then be carried fifty feet with, say six to nine inches rise, and still have sufficient draft. Inside of the propagating room we have again two compartments--the propagating bench, nearest to the furnace, and a shelf for the reception of the young plants, after their first transplanting from the cutting-pots or boxes. Make a shelf or table along the whole length of the house; at the lower end it should be about eighteen inches from the glass, and five feet wide. To a house of, say fifty feet, the propagating bench may be, say twelve feet long, and the room below it and around the flue should be inclosed with boards, as it will keep the heat better.

MODE OF OPERATING.

The wood should be cut from the vines in the fall, as soon as the leaves have dropped. For propagating, use only firm, well-ripened wood of the last season's growth, and about medium thickness. These are to be preferred to either very large or very small ones. The time to commence operating will vary according to climate; here it should be the early part of February. The wood to be used for propagating can be kept in a cool cellar, in sand, or buried in the ground out doors. Take out the cuttings, and cut them up into pieces as represented in Figure 1.

Throw these into water as they are cut; it will prevent them from becoming dry. It will be found of benefit with hard-wooded varieties to pack them in damp moss for a week or so before they are put into the propagating pots or boxes; it will soften the alburnous matter, and make them strike root more readily. They should then be put into, say six-inch pots, filled to about an inch of the top with pure coarse sand, firmly packed. Place the cuttings, the buds up, about an inch apart, all over the surface of the pot; press down firmly with thumb and forefinger until the bud is even with the surface; sift on sand enough to cover the upper point of the bud about a quarter of an inch deep; press down evenly, using the bottom of another pot for the purpose, and apply water enough to moisten the whole contents of the pot. Instead of the pots, shallow boxes of about six inches deep, can also be used, with a few holes bored in the bottom for drainage.

After the pots have been filled with cuttings they are placed in a temperature of from 40?to 45? where they remain from two to three weeks, water being applied only enough to keep them moist, not wet. As roots are formed at a much lower degree of temperature than leaves, they should not be forced too much at the beginning, or the leaves will appear before we have any roots to support them. But when the cutting has formed its roots first, the foliage, when it does appear, will grow much more rapidly, and without any check. Then remove them to another position, plunging the pots into sand to the depth of, say three inches, and raise the temperature at first to 60?for the first few days, then gradually raise it to 80? When the buds begin to push, raise the temperature to 90?or 95? and keep the air moist by frequent waterings, say once a day. The best for this purpose is pure rain-water, but it should be of nearly the same temperature as the air in the house, for, if applied cold, it would surely check the growth of the plants. The young growth should be examined every day, to see if there is any sign of rotting; should this be the case, give a little more air, but admit no sudden cold currents, as they are often fatal. The glass should be whitewashed, to avoid the direct rays of the sun.

When the young vines have made a growth of two or three inches shift them into three-inch pots.

So far we have used only pure sand, which did not contain much plant food, because the growth was produced from the food stored up in the bud and wood, and what little they obtained from the sand, water, and air. Now, however, our young vines want more substantial food. They should therefore be potted into soil, mixed from rotten sod, leaf-mould, and well-decomposed old barnyard manure. This should be mixed together six months before using; add, before using, one-quarter sand, then mix thoroughly, and sift all through a coarse sieve. In operating, put a quantity of soil on the potting bench, provide a quantity of broken bricks or potsherds for drainage, loosen the plants from the pots by laying them on their side, giving them a sudden jar with the hand, to loosen the sand around them; draw out the plant carefully, holding it with one hand, while with the other you place a piece of the drainage material into the pot; cover it with soil about an inch; then put in the plant, holding it so that the roots spread out naturally; fill in soil around them until the pot is full; press the soil down firmly, but not hard enough to break the roots. When the plants are potted give them water to settle the earth around the roots, and keep the air somewhat confined for a few days, until they have become established, when more air may be given them. Keep the temperature at 85?to 95?during the day, and 70?to 80?during the night.

When the plants have made about six inches of growth they can either be placed in another house, or in hot-bed frames, if they are to be kept under glass. The usual manner of keeping them in pots during summer, shifting them into larger and larger sizes, I consider injurious to the free development of the plants, as the roots are distorted and cramped against the sides of the pots, and cannot spread naturally. I prefer shifting them into cold frames, in which beds have been prepared of light, rich soil, into which the young plants can be planted, and kept under whitewashed hot-bed sashes for a while, which, after several weeks, may be removed, and only a light shading substituted in their place, which, after several weeks more, can also be removed. Thus the young plants are gradually hardened, their roots have a chance to spread evenly and naturally, without any cramping; and such plants, although they may not make as tall a growth as those kept under glass all the season, will really stand transplanting into the vineyard much better than those hot-house pets, which may look well enough, but really are, like spoiled and pampered children, but poorly fitted to stand the rough vicissitudes of every-day life.

The young plants should be lightly tied to small sticks provided for the purpose, as it will allow free circulation of air, and admit the sun more freely to the roots. In the fall, after their leaves have dropped, they should be carefully taken up, shortened to about a foot of their growth, and they are then ready either to sell, if they are to be disposed of in that way, or for planting into the vineyard. They should, however, be carefully assorted, making three classes of them--the strongest, medium, and the smallest--each to be put separate. The latter generally are not fit to transplant into the vineyard, but they may be heeled in, and grown in beds another year, when they will often make very good plants. Heeling in may be done as shown in Figure 2, laying the vines as close in the rows as they can conveniently be laid, and then fill the trench with well-pulverized soil. They can thus be safely kept during the winter.

I have only given an outline of the most simple and cheapest mode of growing plants from single eyes, such as even the vineyardist may follow. For descriptions of more extensive and costly buildings, if they desire them, they had better apply to an architect. I have also not given the mode of propagating from green wood, as I do not think, plants thus propagated are desirable. They are apt to be feeble and diseased, and I think, the country at large would be much better off, had not a single plant ever been produced by that method.

Plants from single eyes may also be grown in a common hot-bed; but as in this the heat can not be as well regulated at will, I think it, upon the whole, not desirable, as the expense of a propagating house on the cheap plan I have indicated, is but very little more, and will certainly in the long run, pay much better. Of course, close attention and careful watching is the first requisite in all the operations.

III.--BY CUTTINGS IN OPEN AIR.

This is certainly the easiest and most simple method for the vineyardist; can be followed successfully with the majority of varieties, which have moderately soft wood, and even a part of the hard wood varieties will generally grow, if managed carefully.

MODE OF OPERATING.

There are several methods, which are followed with more or less success. I will first describe that which I have found most successful, namely, short cuttings, of two or three eyes each, which are made of any sound, well ripened wood, of last season's growth. Prune the vines in the fall or early winter, and make the cuttings as soon as convenient; for if the wood is not kept perfectly fresh and green, the cuttings will fail to grow. Now, cut up all the sound, well-ripened wood into lengths of from two to four eyes each, making them of a uniform length of say eight inches, and prepare them as shown in Figure 3.

These should be tied into convenient bundles, from 100 to 250 in each, taking care to even the lower ends, and then buried in the ground, making a hole somewhat deeper than the cuttings are long, into which the bundles are set on their lower ends, and soil thrown in between and over them. In spring, as soon as the ground is dry enough, the cutting-bed should be prepared. Choose for this a light, rich soil, which should be well pulverized, to the depth of at least a foot, and if not light enough, it should be made so by adding some leaf mould. Now draw a line along the whole length of the bed; then take a spade and put it down perpendicular along the line or nearly so, moving it a little backwards and forwards, so as to open the cut. Now take the cutting and press it down into the cut thus made, until the upper bud is even with the surface of the soil. The cuttings may be put close in the rows, say an inch apart, and the rows made two feet apart. Press the ground firmly down with your foot along the line of cuttings, so as to pack it closely around the cutting. After the bed is finished, mulch them with straw, or litter, spent tan or saw-dust, say about an inch thick, and if none of these can be had, leaves from the forest may be used for the purpose. This will serve to protect the young leaves from the sun, and will also keep an even moisture during the heat of summer, at the same time keeping the soil loose and porous. If weeds appear, they should be pulled up, and the cuttings, kept clean through the summer. They will generally make a firm, hardy growth of from one to four feet, have become used to all the hardships and changes of the weather; and as they have formed their roots just where they ought to be, about eight inches below the ground, will not suffer so much from transplanting, as either a single eye or a layer, whose roots have to be put much deeper in transplanting, than they were before, and thus, as it were, become acclimated to the lower regions. For these reasons, I think, that a good plant grown from a cutting is preferable to that propagated by any other method. In the Fall, the vines are carefully taken up, assorted and heeled in, in the same manner as described, with single eyes, and cut back to about three inches of their growth. They are then ready for transplanting into the vineyard.

IV.--BY LAYERING.

This is a very convenient method of increasing such varieties as will not grow readily from cuttings; and vines thus propagated will, if treated right, make very good plants. To layer a vine, shorten in its last season's growth to about one-half; then prepare the ground thoroughly, pulverizing it well; then, early in spring make a small furrow, about an inch deep, then bend the cane down and fasten it firmly in the bottom of the trench, by wooden hooks or pegs, made for the purpose. They may thus be left, until the young shoots have grown, say six inches; then fill up with finely pulverized soil or leaf-mould. The vines will thus strike root generally at every joint. The young shoots may be tied to small sticks, provided for the purpose, and when they have grown about a foot, their tips should be pinched off to make them grow more stocky. In the Fall they are taken up carefully, commencing to dig at the end furthest removed from the vine, and separate each plant between the joints, so that every shoot has a system of roots by itself. They are then either planted immediately, or heeled in as described before.

V.--BY GRAFTING.

The principal advantages to be gained by this method are: 1st. The facility by which new and rare kinds may be increased, by grafting them on strong stocks of healthy varieties, when they will often grow from ten to twenty feet the first season, producing an abundance of wood to propagate. 2d. The short time in which fruit can be obtained from new and untried varieties, as their grafts will generally bear the next season. 3d. In every vineyard there are, in these days of many varieties, vines which have proved inferior, yet by grafting into them some superior variety, they may be made very valuable. 4th. The facility by which vines can be forced under glass, by grafting on small pieces of roots, and the certainty with which every bud can thus be made to grow.

The vine, however, does not unite with the same facility as the pear and apple, and, to ensure success, must be grafted under ground, which makes the operation a difficult and disagreeable one. It will therefore hardly become a general practice; but, for the purposes above named, is of sufficient importance, to make it desirable that every vineyardist should be able to perform it. I have generally had the best success in grafting here about the middle of March, in the following manner: Dig away the ground around the vine you wish to graft, until you come to a smooth place to insert your scion; then cut off the vine with a sharp knife, and insert one or two scions, as in common cleft-grafting, taking care to cut the wedge on the scion very thin, with shoulders on both sides, as shown in Figure 4, cutting your scion to two eyes, to better insure success. Great care must be taken to insert the scion properly, as the inner bark or liber of the vine is very thin, and the success of the operation depends upon a perfect junction of the stock and scion. If the vine is strong enough to hold the scion firmly, no further bandage is necessary; if not, it should be wound firmly and evenly with bass bark. Then press the soil firmly on the cut, and fill up the hole with well pulverized earth, to the top of the scion. Examine the stock from time to time, and remove all wild shoots and suckers, which it may throw up, as they will rob the graft of nourishment and enfeeble it.

Others prefer to graft in May, when the leaves have expanded, and the most rapid flow of sap has ceased, keeping the scions in a cool place, to prevent the buds from starting. The operation is performed in precisely the same manner, and will be just as successful, I think, but the grafts that have been put in early, have the advantage of several weeks over the others, and the latter will seldom make as strong a growth, or ripen their wood as well as those put in early.

Mr. A. S. FULLER performs the operation in the fall, preventing the graft from freezing by inverting a flower-pot over it, and then covering with straw or litter. He claims for this method--1st. That it can be performed at a time when the ground is more dry, and in better condition, and business not so pressing as in spring.--2d. That the scion and stock have more time to unite, and will form their junction completely during the winter, and will therefore start sooner, and make a more rapid growth than in spring. It certainly looks feasible enough, and is well worth trying, as, when the operation succeeds, it must evidently have advantages over any of the other modes.

Vines I had grafted in March have sometimes made twenty to thirty feet of growth, and produced a full crop the next season. This will show one the advantage to be derived from it in propagating new and scarce varieties, and in hastening the fruiting of them. Should a seedling, for instance, look very promising in foliage and general appearance, fruit may be obtained from it from one to two seasons sooner by grafting some of the wood on strong stocks, than from the original plant. Hence the vast importance of grafting, even to the practical vineyardist.

THE VINEYARD.

LOCATION AND SOIL.

As the selection of a proper location is of vast importance, and one of the main conditions of success, great care and judgment should be exercised in the choice. Some varieties of grapes may be grown on almost any soil, it is true; but even they will show a vast difference in the quality of the fruit, even if the quantity were satisfactory; on indifferent soil, and in an inferior location. Everybody should grow grapes enough for his own use, who owns an acre of ground, but every one cannot grow them and make the most delicious wine.

The best locations are generally on the hillsides, along our larger rivers, water-courses, and lakes, sloping to the East, South, and Southwest, as they are generally more exempt from late spring frosts and early frosts in fall. The location should be sheltered from the cold winds from the north and northwest, but fully exposed to the prevailing winds in summer from the south and southwest. If a hill is chosen at any distance from a large body of water, it should be high and airy, with as gentle a slope as can be obtained. The locations along creeks and smaller water-courses should be particularly avoided, as they are subject to late spring frosts, and are generally damp and moist.

The soil should be a dry, calcareous loam, sufficiently deep, say three feet; if possible, draining itself readily. Should this not be the case naturally, it should be done with tiles.

I was much struck by the force of a remark made by medical friend last summer, when, in consequence of the continual rains, the ague was very prevalent. It was this: wherever you will find the ague an habitual guest with the inhabitants you need not look for healthy grapevines. Wherever we find stagnant water let us avoid the neighboring hillsides, for they would not be congenial to our grape-vines. But on the bluffs overhanging the banks of our large streams, especially on the northern and western sides, where the vines are sheltered from the north and west winds, and fully exposed to the warm southern winds of our summer days, and where the fogs arising from the water yet give sufficient humidity to the atmosphere, even in the hottest summer days, to refresh the leaf during the night and morning hours; where the soil on the southern and eastern slopes is a mixture of decomposed stone and leaf-mould, and feels like velvet to the feet--there is the paradise for the grape; and the soil is already better prepared for it than the hand of man can ever do. Such locations should be cheap to the grape-grower at any price. We find them very frequently along the northern banks of the Missouri and Mississippi rivers, and they will no doubt become the favored grape regions of the country. The grape grows there with a luxuriance and health which is almost incredible to those living in less favored locations.

But the question may be asked here, what shall be done by those who do not live in these favored regions, and yet would like to grow grapes? I answer, let them choose the best location they have, the most free and airy, and let them choose only those sturdy varieties that withstand everything. They cannot grow the most delicate varieties--the Herbemont, the Delaware, the Clara, are not for them; but they can grow the Concord, Hartford Prolific, and Norton's Virginia, and they at least are "very good," although they may not be the "best." There is no excuse for any one in this country why he should not grow his own grapes, for the use of his family at least, if he has any ground to grow them on.

PREPARING THE SOIL.

In this, the foundation of all grape-growing, the vineyardist must also look to the condition in which he finds the soil. Should it be free of stones, stumps, and other obstructions, the plough and sub-soil plough will be all-sufficient.

Should your soil be new, perhaps a piece of wild forest land, have it carefully grubbed, and every tree and stump taken out by the roots. After the ground is cleared take a large breaking-plough, with three yoke of sturdy oxen, and plough as deep as you can, say twelve to fourteen inches. Now follow in the same furrow with an implement we call here a sub-soil stirrer, and which is simply a plough-share of wedge shape, running in the bottom of the furrow, and a strong coulter, running up from it through the beam of the plough, sharp in front, to cut the roots; the depth of the furrow is regulated by a movable wheel running in front, which can be set by a screw. With two yoke of oxen this will loosen the soil to the depth of, say twenty inches, which is sufficient, unless the sub-soil is very tenacious. In land already cultivated, where there are no roots to obstruct, two yoke of oxen or four horses attached to the plough, and one yoke of oxen or a pair of horses or mules to the sub-soil plough, will be sufficient. In stony soil the pick and shovel must take the place of the plough, as it would be impossible to work it thoroughly with the latter; but I think there is no advantage in the common method of trenching or inverting the soil, as is now practiced to a very great extent. If we examine the growth of our native vines we will generally find their roots extending along the surface of the soil. It is unnatural to suppose that the grape, the most sun-loving of all our plants, should be buried with its roots several feet below the surface of the soil, far beyond the reach of sun and air. Therefore, if you can afford it, work your soil deep and thoroughly; it will be labor well invested; is the best preventive against drouth, and also the best drainage in wet weather; but have it in its natural position--not invert it; and do not plant too deep. Should the soil be very poor it may be enriched by manure, ashes, bone-dust, etc.; but it will seldom be found necessary, as most of our soil is rich enough; and it is not advisable to stimulate the growth too much, as it will be rank and unhealthy, and injurious to the quality and flavor of the fruit.

Wet spots may be drained by gutters filled with loose stones, or tiles, and then covered with earth. Surface-draining can be done by running a small ditch or furrow every sixth or eighth row, parallel with the hillside, and leading into a main ditch at the end or the middle of the vineyard. Steep hillsides should be terraced or benched; but, as this is very expensive, they should be avoided.

WHAT SHALL WE PLANT?

CHOICE OF VARIETIES.

It is a very difficult matter, in a vast country like ours, where the soil and climate differ so much, to recommend any thing; and I think it a mistake, into which many of our prominent grape-growers have fallen, to recommend any variety, simply because it succeeded well with them, for general cultivation. Grape-growing is, perhaps, more than any other branch of horticulture or pomology, dependent upon soil, location and climate, and it will not do to dictate to the inhabitants of a country, in which the "extremes meet," that they should all plant one variety. Yet this has been done by some who pretend to be authorities, and it shows, more than any thing else, that they have more arrogance than knowledge. I, for my part, have seen such widely different results, from the same varieties, under the same treatment, and in vineyards only a few miles apart, but with a different soil and different aspect, that I am reluctant to recommend to my next neighbor, what he shall plant.

But, while the task is a difficult one, yet we may lay down certain rules, which can govern us in selection of varieties to a certain extent. We should choose--1st. The variety which has given the most general satisfaction in the State or county in which we live, or the nearest locality to us. 2d--Visit the nearest accessible vineyard in the month of August and September, observe closely which variety has the healthiest foliage and fruit; ripens the most uniformly and perfectly; and either sells best in market, or makes the best wine, and which, at the same time, is of good quality, and productive enough. Your observations, thus taken, will be a better guide than the opinion of the most skillful grape grower a thousand miles off.

I will now name a few of the most prominent varieties which should at least be tried by every grape grower.

THE CONCORD.

This grape seems to have given the most general satisfaction all over the country, and seems to be the "grape for the million." Wherever heard from, it seems to be uniformly healthy and productive. Our Eastern friends complain of its inferior quality; this may be owing partly to their short seasons, and partly to the too early gathering of the fruit. It is one of those varieties which color early, but should hang a long time after coloring, to attain its full perfection. Here it is at least very good; makes an excellent wine, and, if we take into consideration its enormous productiveness, its vigor and adaptability to all soils and climates, we must acknowledge that as yet it stands without a rival, and will be a safe investment almost anywhere. Our long summers bring it to a perfection of which our Eastern friends have no idea, until they try it here. It will do well in almost any soil.

NORTON'S VIRGINIA.

This, so far, is the leading grape for red wine, and its reputation here and in the entire West is now so fully established, that it would be difficult indeed to persuade our people into the belief, that any other grape could make a better red wine. It is healthy and uniformly productive, and will be safe to plant, I think, in nearly all the Western States. I rather doubt that our Eastern friends will succeed in making a first class wine from it, as I think their summers are too short, to develop all its good qualities. Will succeed in almost any soil, but attains its greatest perfection in southern slopes with somewhat strong soil.

HERBEMONT.

This is a truly delicious grape, but somewhat tender, and wants a long season to fully ripen its fruit and bring out all its good qualities. Will hardly do much further north than we are here, in Missouri, but is, I think, destined to be one of the leading grapes for the Southern States. If you have a warm, southern exposure, somewhat stony, with limestone foundation, plant the Herbemont, and you will not be disappointed. It is healthy and very productive; more refreshing than the Delaware, and makes an excellent wine.

DELAWARE.

Is much recommended by Eastern authorities, and where it succeeds, is certainly a fine grape and makes a delicious wine. Here at the West, it has proved a failure in most locations, being subject to leaf-blight, and a feeble grower. There are some locations, however, where it will flourish; and whoever is the fortunate possessor of such a one should not forget to plant it. It seems to flourish best in light, warm, somewhat sandy soil.

HARTFORD PROLIFIC.

This is immensely productive; of very fair quality here; hardy and healthy; and if planted for early marketing, will give general satisfaction. It hangs well to the bunch, and even makes a very fair wine. Will flourish in almost every soil.

CLINTON.

Hardy, healthy and productive; will make a fair wine, but is here not equal even to the Concord, and far behind the Norton's Virginia in quality. May be desirable further north.

PLANTING.

The distance at which the vines may be planted will of course vary somewhat with the growth of the different varieties. The rows may all be six feet apart, as this is the most convenient distance for cultivating, and gives ample space for a horse and man to pass through with plough or cultivator. Slow-growing varieties, such as the Delaware and Catawba, may be planted six feet apart in the rows, making the distance six feet each way; but the Concord, Norton's Virginia, Herbemont, Hartford Prolific, Cunningham, and all the strong growers, will need more room, say ten feet in the rows, so as to give the vines ample room to spread, and allow free circulation of air--one of the first conditions of health in the vines, and quality of the fruit.

The next question to be considered is: Shall we plant cuttings or rooted plants? My preference is decidedly for the latter, for the following reasons: Cuttings are uncertain, even of those varieties which grow the most readily; and we cannot expect to have anything like an even growth, such as we can have if the plants are carefully assorted. Some of the cuttings will always fail, and there will be gaps and vacancies which are hard to fill, even if the strongest plants are taken for replanting. Therefore, let us choose plants.

But we should not only choose rooted plants, but the best we can get; and these are good one year old, whether grown from cuttings, layers or single eyes. A good plant should have plenty of strong, well-ripened roots; not covered with excrescences and warts, which is always a sign of ill health; but smooth and firm; with well-ripened, short-jointed wood. They should be of uniform size, as they will then make an even stand in the vineyard, when not forced by the propagator into an unnaturally rank growth by artificial manures. This latter consideration, I think, is very important, as we can hardly expect such plants, which have been petted and pampered, and fed on rich diet, to thrive on the every-day fare they will find in the vineyard. Do not take second or third rate plants, if you can help it; they may live and grow, but they will never make the growth which a plant of better quality would make. We may hear of good results sometimes, obtained by planting second-rate plants, but certainly the results would be better if better plants had been chosen. Especially important is the selection of good plants with those varieties which do not propagate and transplant readily, such as the Norton's Virginia, Delaware, and other hard-wood varieties. Better pay double the price you would have to give for inferior plants; the best are the cheapest in the end, as they will make the healthiest vines, and bear sooner.

But I would also caution my readers against those who will sell you "extra large layers, for immediate bearing," and whose "plants are better than those whom anybody else may grow," as their advertisements will term it. It is time that this humbug should cease; time that the public in general should know, that they cannot, in nature and reason, expect any fruit from a plant transplanted the same season; and that those who pretend it can be done, without vital injury to the plant, are only seeking to fill their pockets at the cost of their customers. They know well enough themselves that it cannot be done without killing or fatally injuring the plant, yet they will impose upon the credulity of their confiding customers; make them pay from $3 to $5 a piece for a plant, which these good souls will buy, with a vision of a fine crop of grapes before their eyes, plant them, with long tops, on which they may obtain a few sickly bunches of fruit the first season; but if they do the vines will make a feeble growth, not ripen their fruit, and perhaps be winter-killed the next season. It is like laying the burden of a full grown man on the shoulders of a child; what was perhaps no burden at all to the one, will kill the other. Then, again, these "plants, superior to those of every one else." It is the duty of every propagator and nursery-man to raise good plants; he can do it if he tries; it is for his interest as much as for the interest of his customers to raise plants of the best quality; and we have no reason to suppose that we are infinitely superior to our neighbors. While the first is a downright swindle, the latter is the height of arrogance. If we had a good deal less of bombast and self laudation, and more of honesty and fair dealing in the profession, the public would have more confidence in professional men, and would be more likely to practice what we preach. Therefore, if you look around for plants, do not go to those who advertise, "layers for immediate bearing," or "plants of superior quality to all others grown;" but go to men who have honesty and modesty enough to send you a sample of their best plants, if required, and who are not averse to let you see how they grow them. Choose their good, strong healthy, one year old plants, with strong, firm, healthy roots, and let those who wish to be humbugged buy the layers for immediate bearing. You must be content to wait until the third year for the first crop; but, then, if you have treated your plants as you ought to do, you can look for a crop that will make your heart glad to see and gather it. You cannot, in reason and nature expect it sooner. If your ground has been prepared in the Fall, so much the better, and if thrown into ridges, so as to elevate the ground somewhat, where the row is to be, they may be planted in the Fall. The advantages of Fall planting are as follows: The ground will generally work better, as we have better weather in the Fall; and generally more time to spare; the ground can settle among the roots; the roots will have healed and callused over, and the young plant be ready to start with full vigor in spring.

Mark your ground, laying it off with a line, and put down a small stick or peg, eighteen inches long, wherever a plant is to stand. Dig a hole, about eight to ten inches deep, as shown in Figure 5, in a slanting direction, raising a small mound in the bottom, of well-pulverized, mellow earth; then, having pruned your plant as shown in Figure 6, with its roots and tops shortened in, as shown by the dotted lines, lay it in, resting the lower end on the mound of earth, spread out its roots evenly to all sides, and then fill in among the roots with rich, well-pulverized earth, the upper bud being left above the ground. When planted in the fall, raise a small mound around your vine, so that the water will drain off, and throw a handful of straw or any other mulch on top, to protect it. Of course, the operation should be performed when the ground is dry enough to be light and mellow, and will readily work in among the roots.

TREATMENT OF THE VINE THE FIRST SUMMER.

The first summer after planting nothing is necessary but to keep the ground free from weeds, and mellow, stirring freely with hoe, rake, plough, and cultivator, whenever necessary. Should the vines grow strong they may be tied to the stakes provided in planting, to elevate them somewhat above the ground. Allow all the laterals to grow, as it will make the wood stronger and more stocky. They may even be summer-layered in July, laying down the young cane, and covering the main stem about an inch deep with mellow soil, leaving the ends of the laterals out of the ground. With free-growing kinds, such as the Concord and Hartford Prolific, these will generally root readily, and make very good plants, the laterals making the stems of the layers. With varieties that do not root so readily, as the Delaware and Norton's Virginia, it will seldom be successful, and should not be practiced. The vineyard may thus be made to pay expenses, and furnish the vines for further plantations the first year. They are taken up and divided in the fall, as directed in the chapter for layers. In the fall, prune the vine to three buds, if strong enough, to one or two if it has only made a weak growth. A fair growth is from four to five feet the first summer. During the winter, trellis should be provided for the vines, as we may expect them to grow from twelve to fifteen feet the coming summer. The cheapest and most economical are those of strong upright posts, say four inches in diameter, made of red cedar if it can be had, if not, of any good, durable timber--mulberry, locust, or white oak--and seven feet long, along which No. 10 wire is stretched horizontally. Make the holes for the posts with a post-hole auger, two feet deep; set in the posts, charred on one end, to make them durable. If wire is to be used, one post every sixteen feet will be enough, with a smaller stake between, to serve as a support for the wires. Now stretch your wire, the lowest one about two feet from the ground, the second one eighteen inches above it, and the third eighteen inches above the second. The wires may be fastened to the posts by nails, around which they can be twisted, or by loops of wire driven into the post. Where timber is plenty, laths made of black oak may be made to serve the same purpose; but the posts must then be set much closer, and the wire will be the cheapest and neatest in the end. A good many grape-growers train their vines to stakes, believing it to be cheaper, but I have found it more expensive than trellis made in the above manner, and it is certainly a very slovenly method, compared with the latter. Trellis is much more convenient for tying the vines, the canes can be distributed much more evenly, and the fruit and young wood, not being huddled and crowded together as on stakes, will ripen much more evenly, and be of better quality, as the air and sun have free access to it.

TREATMENT OF THE VINE THE SECOND SUMMER.

We find the young vine at the commencement of this season pruned to three buds of the last season's growth. From these we may expect from two to three strong shoots or canes. Our first work will be to cultivate the whole ground, say from four to six inches deep, ploughing between the rows, and hoeing around the vines with a two-pronged German hoe, or karst. Figure 7 shows one of these implements, of the best form for that purpose. The ground should be completely inverted, but never do it in wet weather, as this will make the ground hard and cloggy.

Of the young shoots, if there are three, leave only the two strongest, tying the best of them neatly to the trellis with bass, or pawpaw bark, or rye straw. If a Catawba or Delaware, you may let them grow unchecked, tying them along the uppermost wire, when they have grown above it. The Concord, Herbemont, Norton's Virginia, and other strong-growing varieties, I treat in the following manner: When the young shoot has reached the second wire I pinch off its leader. This has the tendency to force the laterals into stronger growth, each forming a medium-sized cane. On these we intend to grow our fruit the coming season, as the buds on these laterals will generally produce more and finer fruit than the buds on the strong canes. Figure 8 will show the manner of training the second summer, with one cane layered, for the purpose of raising plants. This is done as described before; only, as the vine will make a much stronger growth this season than the first, the layering maybe done in June, as soon as the young shoots are strong enough. Figure 9 shows the vine pruned and tied, at the end of the second season. Figure 10 illustrates the manner of training and tying the Catawba or Delaware.

The above is a combination of the single cane and bow system, and the horizontal arm training, which I first tried on the Concord from sheer necessity; when the results pleased me so much that I have adopted it with all strong-growing varieties. The circumstances which led me to the trial of this method were as follows: In the summer of 1862, when my Concord vines were making their second season's growth, we had, in the beginning of June, the most destructive hail storm I have ever seen here. Every leaf was cut from the vines, and the young succulent shoots were all cut off to about three to three and a half feet above the ground. The vines, being young and vigorous, pushed out the laterals vigorously, each of them making a fair-sized cane. In the fall, when I came to prune them, the main cane was not long enough, and I merely shortened in the laterals to from four to six buds each. On these I had as fine a crop of grapes as I ever saw, fine, large, well-developed bunches and berries, and a great many of them, as each had produced its fruit-bearing shoot. Since that time I have followed this method altogether, and obtained the most satisfactory results.

The ground should be kept even and mellow during the summer, and the vines neatly tied to the trellis with bast or straw.

There are many other methods of training; for instance, the old bow and stake training, which is followed to a great extent around Cincinnati, and was followed to some extent here. But it crowds the whole mass of fruit and leaves together so closely that mildew and rot will follow almost as a natural consequence, and those who follow it are almost ready to give up grape-culture in despair. Nor is this surprising. With their tenacious adherence to so fickle a variety as the Catawba, and to practices and methods of which experience ought to have taught them the utter impracticability long ago, we need not be surprised that grape-culture is with them a failure. We have a class of grape-growers who never learn, nor ever forget, anything; these we cannot expect should prosper. The grape-grower, of all others, should be a close observer of nature in her various moods, a thinking and a reasoning being; he should be trying and experimenting all the time, and be ready always to throw aside his old methods, should he find that another will more fully meet the wants of his plants. Only thus can he expect to prosper.

There is also the arm system, of which we hear so much now-a-days, and which certainly looks very pretty on paper. But paper is patient, and while it cannot be denied that it has its advantages, if every spur and shoot could be made to grow just as represented in drawings, with three fine bunches to each shoot; yet, upon applying it practically, we find that vines are stubborn, and some shoots will outgrow others; and before we hardly know how, the whole beautiful system is out of order. It may do to follow in gardens, on arbors and walls, with a few vines, but I do not think that it will ever be successfully followed in vineyard culture for a number of years, as it involves too much labor in tying up, pruning, etc. I think the method described above will more fully meet the wants of the vinyardist than any I have yet seen tried; it is so simple that every intelligent person can soon become familiar with it, and it gives us new, healthy wood for bearing every season. Pruning may be done in the fall, as soon as the leaves have dropped.

TREATMENT OF THE VINE THE THIRD SEASON.

At the commencement of the third season, we find our vine pruned to two spurs of two eyes each, and four lateral canes, of from four to six eyes each. These are tied firmly to the trellis as shown in Figure 12, for which purpose small twigs of willows (especially the golden willow, of which every grape-grower should plant a supply) are the most convenient. The ground is ploughed and hoed deeply, as described before, taking care, however, not to plough so deep as to cut or tear the roots of the vine.

Our vines being tied, ploughed, and hoed, we come to one of the most important and delicate operations to be performed; one of as great--nay, greater--importance than pruning. I mean summer-pruning, or pinching, i.e. thumb or finger pruning. Fall-pruning, or cutting back, is but the beginning of the discipline under which we intend to keep our vines; summer-pruning is the continuation, and one is useless, and cannot be followed systematically without the other.

Let us look at our vine well, before we begin, and commence near the ground. The time to perform the first summer-pruning is when the young shoots are about six to eight inches long, and when you can see plainly all the small bunches or buttons--the embryo fruit. We commence on the lower two spurs, having two buds each. From these two shoots have started. One of them we intend for a bearing cane next summer; therefore allow it to grow unchecked for the present, tying it, if long enough, to the lowest wire. The other, which we intend for a spur again next fall, we pinch with thumb and finger to just beyond the last bunch or button, taking out the leader between the last bunch and the next leaf, as shown in Figure 11, the cross line indicating where the leader is to be pinched off. We now come to the next spur, on the opposite side, where we also leave one cane to grow unchecked, and pinch off the other. We now go over all the shoots coming from the arms or laterals tied to the trellis, and also pinch them beyond the last bunch. Should any of the buds have pushed out two shoots, we rub off the weakest; we also take off all barren or weak shoots. If any of them are not sufficiently developed we pass them over, and go over the vines again, in a few days after the first pinching.

This early pinching of the shoot has a tendency to throw all the vigor into the development of the young bunch, and the leaves remaining on the shoot, which now grow with astonishing rapidity. It is a gentle checking, and leading the sap into other channels; not the violent process which is often followed long after the bloom, when the wood has become so hardened that it must be cut with a knife, and by which the plant is robbed of a large quantity of its leaves, to the injury of both fruit and vine. Let any of my readers, who wish to satisfy themselves, summer-prune a vine, according to the method described here, and leave the next vine until after the bloom, and he will plainly perceive the difference. The merit of first having practised this method here, which I consider one of vast importance in grape-culture, belongs to Mr. WILLIAM POESCHEL, of this place, who was led to do so, by observing the rapid development of the young bunches on a shoot which had accidentally been broken beyond the last bunch. Now, there is hardly an intelligent grape-grower here, who does not follow it; and I think it has added more than one-third to the quantity and quality of my crop. It also gives a chance to destroy the small, white worm, a species of leaf-folder, which is very troublesome just at this time, eating the young fruit and leaves, and which makes its web among the tender leaves at the end of the shoot.

The bearing branches having all been pinched back, we can leave our vines alone until after the bloom, only tying up the young canes from the spurs, should it become necessary. But do not tie them over the bearing canes, but lead them to the empty space on both sides of the vine; as our object must be to give the fruit all the air and light we can.

By the time the grapes have bloomed, the laterals will have pushed from the axils of the leaves on the bearing shoots. Now go over these again, and pinch each lateral back to one leaf, as shown in Figure 12. This will make the leaf which remains grow and expand rapidly, serving at the same time as a conductor of sap to the young bunch opposite, and shading it when it becomes fully developed. The canes from the spurs, which we left unchecked, and which we design to bear fruit the next season, may now also be stopped or pinched, when they are about three feet long, to start their laterals into stronger growth. Pinch off all the tendrils; this is a very busy time for the vine-dresser, and upon his close attendance and diligence now, depends, in a great measure, the value of his crop. Besides, "a stitch in time saves nine," and he can save an incredible amount of labor by doing everything at the proper time.

In a short time, the laterals on the fruit-bearing branches which have been pinched will throw out suckers again. These are stopped again, leaving one leaf of the young growth. Leave the laterals on the canes intended for next years' fruiting to grow unchecked, tying them neatly with bass, or pawpaw bark, or with rye straw.

This is about all that is necessary for this summer, except an occasional tying up of a fruiting branch, should its burden become more than it can bear. But the majority of the branches will be able to sustain their fruit without tying, and the young growth which may yet start from the laterals may be left unchecked, as it will serve to shade the fruit when ripening. Of course, the soil must be kept clean and mellow, as in the former summer. This short pruning is also a partial preventative against mildew and rot, and the last extremely wet season has again shown the importance of letting in light and air to all parts of the vine; as those vineyards, where a strict system of early summer pruning had been followed, did not suffer half as much from rot and mildew as those where the old slovenly method still prevailed.

My readers will perceive, that Fall-pruning, or shortening-in the ripened wood of the vine, and summer-pruning, shortening in and thinning out the young growth, have one and all the same object in view, namely, to keep the vine within proper bounds, and concentrate all its energies for a two-fold object, namely, the production and ripening of the most perfect fruit, and the production of strong, healthy wood for the coming season's crop. Both operations are, in fact, only different parts of one and the same system, of which summer-pruning is the preparatory, and fall pruning the finishing part.

If we think that a vine is setting more fruit than it is able to bear and ripen perfectly, we have it in our power to thin it, by taking away all imperfect bunches, and feeble shoots. We should allow no more wood to grow than we need for next season's bearing; if we allow three canes to grow where only two are needed, we waste the energies of the vine, which should all be concentrated upon ripening its fruit in the most perfect condition, and producing the necessary wood for next season's bearing, and that of the best and most vigorous quality, but no more. If we prune the vine too long, we over-tax its energies; making it bear more fruit than it can perfect, and the result will be poor, badly-ripened fruit, and small and imperfect wood. If, on the contrary, we prune the vine too short, we will have a rank, excessive growth of wood and leaves, and encourage rot and mildew. Only practice and experience will teach us the exact medium, and the observing vintner will soon find out where he has been wrong, better than he can be taught by a hundred pages of elaborate advice. Different varieties will require different treatment, and it would be foolishness to suppose that two varieties so entirely different, as for instance, the Concord and the Delaware, could be pruned, trained and pinched in the same manner. The first, being a rank and vigorous grower, with long joints, will require much longer pruning than the latter, which is a slow-growing, short-jointed vine. Some varieties, the Taylor for instance, also the Norton, will fruit better if pruned to spurs on old wood, than on the young canes; it will therefore be the best policy for the vintner in pruning these, to retain the old arms or canes, pruning all the healthy, strong shoots they have to two buds, as long as the old arms remain healthy; always, however, growing a young cane to fall back upon, should the old arm become diseased; whereas, the Catawba and Delaware, being only moderate growers, will flourish and bear best when pruned short, and to a cane of last season's growth. The Concord and Herbemont, again, will bear best on the laterals of last season's growth, and should be trained accordingly. Therefore it is, because only a few of the common laborers will take the pains to think and observe closely, that we find among them but few good vine-dressers.

At the end of this season, we find our Concords or Herbemonts, with the old fruit-bearing cane, and a spur on each side, from which have grown two canes; one of which was stopped, like all other fruit-bearing branches, and which we now prune to a spur of two eyes; and another, which was stopped at about three feet, and on which the laterals were allowed to grow unchecked. We therefore have one of these canes, with its laterals, on each side of the vine. These laterals are now pruned precisely as the last season, each being cut back to from four to six eyes, and the old cane, which has borne fruit, is cut away altogether. With Norton's Virginia, Taylor, and some others, which will bear more readily on spurs from old wood, the old cane is retained, provided the shoots on it are sound and healthy, with well developed buds; the weak ones are cut away altogether, and the others cut back to two eyes each. One of the canes is pruned, as in the Concord, to be tied to one side of the trellis, the next spring. This closes our summer and fall pruning for the third year. Of the gathering of the fruit, as well for market as for wine, I shall speak in another chapter.

TREATMENT OF THE VINE THE FOURTH SUMMER.

We may now consider the vine as established, able to bear a full crop, and when tied to the trellis in spring, to present the appearance, as shown in Fig. 13. The operations to be performed are precisely the same as in its third year.

In addition, I will here remark, that in wet seasons the soil of the vineyard should be stirred as little as possible, as it will bake and clog, and in dry seasons it should be deeply worked and stirred, as this loose surface-soil will retain moisture much better than a hard surface. Should the vines show a decrease in vigor, they may be manured with ashes or compost, or still better, with surface-soil from the woods. This will serve to replenish the soil which may have been washed off and is much more beneficial than stable manure. When the latter is applied, a small trench should be dug just above the vine, the manure laid in, and covered with soil. But an abundance of fresh soil, drawn up well around the vine, is certainly the best of all manures.

Where a vine has failed to grow the first season, replant with extra strong vines, as they will find it difficult to catch up with the others; or the vacancy can be filled up the next season, by a layer from a neighboring vine, made in the following manner: Dig a trench from the vine to the empty place, about eight to ten inches deep, and bend into it one of the canes of the vine, left to grow unchecked for that purpose, and pruned to the proper length. Let the end of it come out to the surface of the ground with one or two eyes above it, at the place where the vine is to be, and fill up with good, well pulverized earth. It will strike roots at almost every joint, and grow rapidly, but, as it takes a good deal of nourishment from the parent vine, that must be pruned much shorter the first year. When the layer has become well established, it is cut from the parent vine; generally the second season.

Pruning is best done in the fall, but it can be done on mild days all through the winter months, even as late as the middle of March. Fall-pruning will prevent all flow of sap, and the cuttings are also better if made in the fall, and buried in the ground during winter. All the sound, well-ripened wood of last season's growth may be made into cuttings, which may be either planted, as directed in a former chapter, or sold; and are an accession to the product of the vineyard not to be despised, for they will generally defray all expenses of cultivation.

TRAINING THE VINES ON ARBORS AND WALLS.

This is altogether different from the treatment in vineyards; the first has for its object to grow the most perfect fruit, and to bring the vine, with all its parts, within the easy reach and control of the operator; in the latter, our object is to cover a large space with foliage, for ornament and shade, fruit being but a secondary consideration. However, if the vine is treated judiciously, it will also produce a large quantity of fruit, although not of as good quality as in the vineyard.

Our first object must be to grow very strong plants, to cover a very large space. Prepare a border by digging a trench two feet deep and four feet wide. Fill with rich soil, decomposed leaves, burnt bones, ashes, etc. Into this plant the strongest plants you have, pruned as for vineyard planting. Leave but one shoot to grow on them during the first summer, which, if properly treated, will get very strong. Cut back to three buds the coming fall. These will each throw out a strong shoot, which should be tied to the arbor they are designed to cover, as shown in Figure 14, and allowed to grow unchecked. In the fall following cut each shoot back to three buds, as our first object must be to get a good basis for our vines. These will give us nine canes the third summer; and as the vine is now thoroughly established and strong, we can begin to work in good earnest. It will be perceived that the vine has three different sections or principal branches, each with three canes. Cut one of these back to two eyes, and the other two to six or eight buds each, according to the strength of the vine, as shown in Figure 15. The next spring tie these neatly to the trellis, and when the young shoots appear thin out the weakest, and leave the others to grow unchecked. The next fall cut back as indicated by the black cross lines, the weakest to be cut back to one or two eyes, and the stronger ones to three or four, the spurs at the bottom to come in as a reserve, should any of the branches become diseased. Figure 16 shows the manner of pruning.

In this manner a vine can be made, in course of time, to cover a large space, and get very old. The great vine at Windsor Palace was planted more than sixty years ago, and in 1850 it produced two thousand large bunches of magnificent grapes. The space covered by the branches was one hundred and thirty-eight feet long, and sixteen feet wide, and it had a stem two feet nine inches in circumference. This is one of the largest vines on record. They should, however, be strongly manured to come to full perfection.

Other authorities prefer the Thomery system of training, but I think it much more complicated and difficult to follow. Those wishing to follow it will find full directions in DR. GRANT'S and FULLER'S books, which are very explicit on this method.

OTHER METHODS OF TRAINING THE VINE.

There are many other systems in vogue among vine-dressers in Germany and France, but as our native grapes are so much stronger in growth, and are in this climate so much more subject to mildew and rot, I think these methods, upon the whole, but poorly adapted to the wants of our native grapes, however judicious they may be there. I will only mention a few of them here; one because it is to a great extent followed in Mexico and California, and seems to suit that dry climate and arid soil very well; and the other, because it will often serve as a pretty border to beds in gardens. The first is the so-called buck or stool method of training. The vine is made to form its head--i.e., the part from which the branches start--about a foot above the ground, and all the young shoots are allowed to grow, but summer-pruned or checked just beyond the last bunch of grapes. The next spring all of the young shoots are cut back to two eyes, and this system of "spurring in" is kept up, and the vine will in time present the appearance of a bush or miniature tree, producing all its fruit within a foot from the head, and without further support than its own stem. Very old vines trained in this manner often have twenty to twenty-five spurs, and present, with their fruit all hanging in masses around the main trunk, a pleasing but rather odd aspect. This method could not be applied here with any chance of success only to those varieties which are slow growers, and at the same time very hardy. The Delaware would perhaps be the most suitable of all varieties I know for a trial of this method; such strong growers as the Concord and Norton's Virginia could never be kept within the proper bounds, and it would be useless to try it on them. It might be of advantage on poor soil, where there is at the same time a scarcity of timber. Figure 17 shows an old vine pruned after this method.

The other method of dwarfing the grape is practiced to make a pretty border along walks in gardens, and is as follows: Plant your vines about eight feet apart; treat them the first season as in common vineyard planting, but at the end of the first season cut back to two eyes. Now provide posts, three to three and a half feet long; drive them into the ground about eighteen inches to two feet, which can be easily done if they are pointed at one end, and nail a lath on top of them. This is your trellis for the vines, and should be about eighteen inches above the ground when ready. Now allow both shoots which will start from the two buds to grow unchecked; and when they have grown above the trellis, tie one down to the right, the other to the left, allowing them to ramble at will along it. The next fall they are each cut back to the proper length, to meet the next vine, and in spring tied firmly to the lath, as shown in Figure 18. When the young shoots appear, all below the trellis are rubbed off, but all those above the trellis are summer-pruned or pinched immediately beyond the last bunch of grapes, as in vineyard culture, and the trellis, with its garland of fruit, will present a very pretty appearance throughout the summer. In the fall all of these shoots are pruned to one bud, from which will grow the fruit-bearing shoot for the next season, as shown in Figure 19; and the same treatment is repeated during the summer and fall.

DISEASES OF THE VINE.

I cannot agree with Mr. FULLER that the diseases of the vine are not formidable in this country. They are so formidable that they threaten to destroy some varieties altogether; and the Catawba, once the glory and pride of the Ohio vineyards, has for the last fifteen years suffered so much from them, that many of the grape-growers who are too narrow-minded to try anything else are about giving up grape-growing in despair.

It is very fortunate, therefore, that we have varieties which do not suffer from these diseases, or only in a very slight degree; and my advice to the beginner in grape-culture would be, "not to plant largely of any variety which is subject to disease." Men may talk about sulphuring, and dusting their vines with sulphur through bellows; but I would rather have vines which will bear a good crop without these windy appliances. We can certainly find some varieties for every locality which do not need them, and these we should plant.

The mildew is our most formidable disease, and will very often sweep away two-thirds of a crop of Catawbas in a few days. It generally appears here from the first to the fifteenth of June, after abundant rains, and damp, warm weather. It seems to be a parasitic fungus, and sulphur applied by means of a bellows, or dusted over the fruit and vine is said to be a partial remedy. Close and early summer-pruning will do much to prevent it, throwing, as it does, all the strength of the vine into the young fruit, developing it rapidly, and also allowing free circulation of air. In some varieties--for instance, the Delaware--it will only affect the leaves, causing them to blight and drop off, after which the fruit, although it may attain full size, will not ripen nor become sweet, but wither and drop off prematurely. In seasons when the weather is dry and the air pure, it will not appear. It is most prevalent in locations which have a tenacious subsoil, and under-draining will very likely prove a partial preventive, as excess of moisture about the roots is no doubt one of its causes.

The gray rot, or so-called grape cholera, generally follows the mildew, and I think that the latter is the principal cause of it, as I have generally found it on berries whose stems have been injured by the mildew. The berry first shows a sort of gray marbling; in a day or two it turns to a grayish-blue color, and finally withers and drops from the bunch. It will continue to affect berries until they begin to color, but only attack a few varieties--the Catawba, To Kalon, Kingsessing, and sometimes the Diana.

The spotted, or brown rot, will also attack many of our varieties; it is very destructive to the Isabella and Catawba, and even the Concord is not quite free from it. But it is, after all, not very destructive, and not half as dangerous as the mildew or gray rot.

Early and close summer-pruning is a partial preventative against all these diseases, as it will hasten the development of the fruit, allow free circulation of air, and the young leaves which appear on the laterals after pinching seem to be better able to withstand the effects of the mildew, often remaining fresh and green, and shading the fruit, when the first growth of leaves have already dropped.

But "an ounce of prevention is better than a pound of cure," and our best preventive is to plant none but healthy varieties. A grape, however good it may be in quality, is not fit for general cultivation if seriously affected with any of these diseases. Nothing can be more discouraging to the grape-grower than to see his vines one day rich in the promise of an abundant crop, and a few days afterwards see two-thirds or three-fourths swept away by disease. It is because I have so often felt this bitter disappointment, that I would warn my readers against planting varieties subject to them. I would save them from the discouragement and bitter losses which I have experienced, when it was out of my power to prevent it. They can prevent it, for the grape-growing of to-day is no longer the same uncertain occupation it was ten years ago. We of to-day have our choice of varieties not subject to disease; let us make it judiciously, and we may be sure of a paying crop every year.

INSECTS INJURIOUS TO THE GRAPE.

The grape has many enemies of this kind, but if they are closely watched from the beginning their ravages are easily kept within proper bounds.

The common gray cut-worm will often eat the young tender shoots of the vine, and draw them into the ground below. Wherever this is perceived the rascal can easily be found by digging for him under some of the loose clods of ground below the vine, and should be destroyed without mercy.

Small worms, belonging to the family of leaf-folders, some of them whitish gray, some bluish green, will in spring make their webs among the young, downy leaves at the end of the shoots, eating the young bunches or buttons, and the leaves. These can be destroyed when summer pruning for the first time. Look close for them, as they are very small; yet very destructive if let alone.

A small, gray beetle, of about the size and color of a hemp-seed, will often eat a hole into the bud, when it is just swelling, and thus destroy it. He is very shy, and will drop from the vine as soon as you come near him. It is a good plan to spread a newspaper under the vine, and then shake it, when he will drop on the paper and can be caught.

Another bug, of about the size of a fly, gray, with round black specks, will sometimes pay us a visit. They will come in swarms, and eat the upper side of the leaves, leaving only the skeletons. They are very destructive, devouring every leaf, as far as they go; they can also be shaken off on a paper or sheet spread under the vine.

The thrip, a small, rather three-cornered, whitish-green insect, has of late been very troublesome, as they eat the under side of the leaves of some varieties, especially the Delaware and Norton's Virginia, when the leaf will show rusty specks on the surface, and finally drop off. It has been recommended to go through the vineyard at night, one man carrying a lighted torch, and the other beating the vines, when they will fly into the flame, and be burnt. They are a great annoyance, and have defoliated whole vineyards here last fall.

Another leaf-folder makes his appearance about mid-summer, making its web on the leaf, drawing it together, and then devouring his own house. It is a small, greenish, and very active worm, who, if he "smells a rat," will drop out of his web, and descend to the ground in double-quick time. I know of no other plan, than to catch him and crush his web between the finger and thumb.

The aphis, or plant louse, often covers the young shoots of the vine, sucking its juices. When a shoot is attacked by them, it will be best to take it off and crush them under your feet, as the shoot is apt to be sickly afterwards, any way.

The grape vine sphynx will be found occasionally. It is a large, green worm, with black dots, and very voracious. Fortunately, it is not numerous, and can easily be found and destroyed.

There are also several caterpillars--the yellow bear, the hog caterpillar, and the blue caterpillar, which will feed upon the leaves. The only remedy I know against them is hand picking, but they have not as yet been very numerous, nor very destructive.

Wasps are sometimes very troublesome when the fruit ripens, stinging the berries and sucking the juice. A great many can be caught by hanging up bottles, with a little molasses, which they will enter, and get stuck in the molasses.

BIRDS.

These are sometimes very troublesome at the time of ripening, and especially the oriole is a "hard customer," as he will generally dip his bill into every berry; often ruining a fine bunch, or a number of them, in a short time. I have therefore been compelled to wage a war upon some of the feathered tribe, although they are my especial favorites, and I cannot see a bird's nest robbed. However, there are some who do not visit the vineyard, except for the purpose of destroying our grapes, and these can not complain if we "won't stand it any longer," but take the gun, and retaliate on them. The oriole, the red bird, thrush, and cat bird are among the number, and although I would like to spare the latter three, in thankful remembrance of many a gratuitous concert, the first must take his chance of powder and lead, for the little rascal is too aggravating. A few dry bushes, raised above the trellis will serve as their resting place before they commence their work of destruction, where they can be easily killed.

FROSTS.

Although our winters are seldom severe enough to destroy the hardy varieties, yet they will often fatally injure such half hardy varieties as the Herbemont and Cunningham, and the severe winter of 1863,-'64, killed even the Catawba, down to the snow line, and severely injured the Norton's Virginia, and even the Concord. Fortunately, such winters occur but rarely, and even in localities where the vines are often destroyed by the severe cold in winter, this should deter no one from growing grapes, as, with very little extra labor he can protect them, and bring them safely through the winter. I always cover my tender varieties, in fact, all that I feel not quite safe to leave out, even in severe winters, in the following manner: The vines are properly pruned in the fall; then select a somewhat rainy day, when the canes will bend more easily. One man goes through the rows, and bends the canes to the ground along the trellis, while another follows with the spade, and throws earth enough on them to hold them in their places. Afterwards, I run a plough through the rows, and cover them up completely. In the spring when all danger from frost is over, I take a so-called spading fork, and lift the vines. The entire cost of covering an acre of grape vines and taking them up again in spring, will not exceed $10; surely a trifling expense, if we can thereby ensure a full crop.

We have thus a protection against the cold in winter, but I know none against early frosts, in fall, and late spring frosts; and the grape grower should therefore avoid all localities where they are prevalent. The immediate neighborhood of large streams, or lakes, will generally save the grape grower from their disastrous influence; and our summers, here, along the banks of the Missouri river, are in reality full two months longer than they are in the low, small valleys, only four to six miles off. Let the grape grower, in choosing a locality, look well to this, and avoid the hills along these narrow valleys. Either choose a location sufficiently elevated, to be beyond their influence, or, what is better still, choose it on the bluffs above our large streams; where the atmosphere, even in the heat of summer, will never become too dry for the health of the vine. It is a sad spectacle to see the hopes of a whole summer frustrated by one cold night; to see the vines which promised an abundant crop but the day before, browned and wilted beyond all hopes of recovery, and the cheerless prospect before you, that it may occur every spring; or to see the finest crop of grapes, when just ripening, scorched and wilted by just one night's frost, fit for nothing but vinegar. Therefore, look well to this, when you choose the site of your vineyard, and rather pay five times the price for a location free from frost, than for the richest farm along the so-called creek bottoms, or worse still, sloughs of stagnant water.

GIRDLING THE VINE TO HASTEN MATURITY.

The practice of girdling to induce early ripening is supposed to have been invented by Col. BUCHATT, of Metz, in 1745. He claimed for it that it would also greatly improve the quality of the fruit, as well as hasten maturity. That it accomplishes the latter, cannot be denied; it also seems to increase the size of the berries, but I hardly think the fruit can compare in flavor with a well developed bunch, ripened in the natural way. As it may be of practical value to those who grow grapes for the market, enabling them to supply their customers a week earlier at least, and also make the fruit look better, and be of interest to the amateur cultivator, I will describe the operation for their benefit.

It can be performed either on wood of the same season's growth, or on that of last year, but in any case only upon such as can be pruned away the next fall. If you desire to affect the fruit of a whole arm or cane, cut away a ring of bark by passing your knife all around it, and making another incision from a quarter to half an inch above the first, taking out the intermediate piece of bark clean, down to the wood. It should be performed immediately after the fruit is set. The bunches of fruit above the incision will become larger, and the fruit ripen and color finely, from a week to ten days before the fruit on the other canes. Of course, the cane thus girdled, cannot be used for the next season, and must be cut away entirely. The result seems to be the consequence of an obstruction to the downward flow of the sap, which then develops the fruit much faster.

Ripening can also be hastened by planting against the south side of a wall or board fence, when the reflection of the rays of the sun will create a greater degree of warmth.

But nothing can be so absurd and unnatural than the practice of some, who will take away the leaves from the fruit, to hasten its ripening. The leaves are the lungs of the plants; the conductors and elevators of sap; and nothing can be more injurious than to take them away from the fruit at the very time when they are most needed. The consequence of such an unwise course will be the wilting and withering of the bunches, and, should they ripen at all, they will be deficient in flavor. Good fruit must ripen in the shade, only thus will it attain its full perfection.

Another practice very injurious to the vines is still in practice in some vineyards, and cannot be too strongly condemned. It is the so-called "cutting in" of the young growth in August. Those who practice it, seem to labor under the misapprehension that the young canes, after they have reached the top of the trellis, and are of the proper length and strength for their next year's crop, do not need that part of the young growth beyond these limits any more, and that all the surplus growth is "of evil." Under the influence of this idea they arm themselves with a villainous looking thing called a bill-hook, and cut and slash away at the young growth unmercifully, taking away one-half of the leaves and young wood at one fell swoop. The consequence is a stagnation of sap: the wood they have left, cannot, and ought not to ripen perfectly, and if anything like a cold winter follows, the vines will either be killed entirely, or very much injured at least. The intelligent vine dresser will tie his young canes, away from the bearing wood as much as he can, to give the fruit the fullest ventilation; but when they have reached the top of the trellis, tie them along it and let them ramble as they please. They will thus form a natural roof over the fruit, keep off all injurious dews, and shade the grapes from above. There is nothing more pleasing to the eye than a vineyard in September, with its wealth of dark green foliage above, and its purple clusters of fruit beneath, coyly peeping from under their leafy covering. Such grapes will have an exquisite bloom, and color, as well as thin skin and rich flavor, which those hanging in the scorching rays of the sun can never attain.

MANURING THE VINE.

As remarked before, this will seldom be necessary, if the vintner is careful enough to guard against washing of the top-soil, and to turn under all leaves, etc., with the plow in the Fall. The best manure is undoubtedly fresh surface soil from the woods. Should the vines, however, show a material decrease in vigor, it may become necessary to use a top-dressing of decomposed leaves, ashes, bone-dust, charcoal, etc. Fresh stable-yard manure I would consider the last, and only to be used when nothing better can be obtained. Turn under with the plow, as soon as the manure is spread. Nothing, I think, is more injurious than the continual drenching with slops, dish-water, etc., which some good souls of housewives are fond of bestowing on their pet grape vines in the garden. It creates a rank, unwholesome growth, and will cause mildew and rot, if anything can.

THINNING OF THE FRUIT.

This will sometimes be necessary, to more fully develop the bunches. The best thinning is the reduction of the number of bunches at the time of the first summer pruning. If a vine shows more fruit, than the vine dresser thinks it can well ripen, take away all weak and imperfect shoots, and also all the small and imperfect bunches. If the number of bunches on the fruit bearing branches is reduced to two on each, it will be no injury, but make the remaining number of bunches so much more perfect. Thinning out the berries on the bunches, although it will serve to make the remaining berries more perfect and larger, is still a very laborious process, and will hardly be followed to any extent in vineyards, although it can well be practised on the few pet vines of the amateur, and will certainly heighten the beauty of the bunches and berries.

RENEWING OLD VINES.

Should a vine become old and feeble, it can be renewed by layering. The vine is prepared in the following manner: Prune all the old wood away, leaving but one of the most vigorous of your canes; then dig a trench from the vine along the trellis, say three feet long, eight inches deep; into this bend down the old vine, stump, head and all, fastening it down with a strong hook, if necessary, letting the end of the young cane come out about three eyes above the ground, and fill up with rich, well pulverized soil. The vine will make new roots at every joint, and become vigorous, and, so to say, young, again. Some recommend this process for young vines, the first year after planting; but if good plants have been chosen and planted, it will not be necessary. Feeble and poor plants may need this process, but if plants have good strong roots when planted, (and only such should be planted when they can be obtained), they will not be benefited by it.

A FEW NECESSARY IMPROVEMENTS.

Pruning Shears. These are very handy, and with them the work can be done quicker, and with less labor, as but a slight pressure of the hand will cut a strong vine. Fig. 22 will show the shape of one for heavy pruning. They are made by J. T. HENRY, Hampden, Connecticut, and can be had in almost all hardware stores. The springs should be of brass, as steel springs are very apt to break. A much lighter and smaller kind, with but one spring, is very convenient for gathering grapes, as it will cut the stem easily and smoothly, and not shake the vine, as cutting with the knife will do. They are also handy to clip out unripe and rotten berries, and should be generally used instead of knives.

Pruning Saws. It will sometimes be necessary to use these, to cut out old stumps, etc., although, if a vine is well managed, it will seldom be necessary. Fig. 23 will show a kind which is very convenient for the purpose, and will also serve for orchard pruning; the blade is narrow, connected with the handle, and can be turned in any direction.

GATHERING THE FRUIT FOR MARKET.

In this, the vineyardist, of course, only aims at profit, and for that purpose the grapes are often gathered when they are hardly colored--long before they are really ripe--because the public will generally buy them at a high price. Let us hope, however, that better taste will in time prevail, and that even a majority of the public will learn to appreciate the difference between ripe and unripe fruit. I would advise my readers at least to wait until the fruit is fully and evenly colored; for it is our duty to do all we can to correct this vicious leaning towards swallowing unripe fruit, which is so prevalent in this nation, and the producer will not lose anything either, because his fruit will look much better, it will therefore bring the same price which half ripened fruit would have brought, even a week sooner, and will weigh heavier. Every grape will generally color full two weeks before it is fully ripe; and as they are one of the fruits that will not ripen after they are gathered, they will shrivel and look indifferent if gathered before.

To ship them to market any distance, they should be packed in low, shallow boxes, say six inches high, so that they will hold about two layers of grapes. Cut the branches carefully, with as long a stem as possible, for more convenient handling, taking care to preserve all the bloom, and clipping out all the unripe berries. They are generally weighed in the basket before packing. Now put a layer of vine leaves on the bottom of the box; then make a layer of grapes, laying them as close as possible; then put a layer of leaves over them; on them put another layer of grapes, filling up evenly; then spread leaves rather thickly over them, and nail on the cover. The box should be perforated with holes, to admit some air. The grapes must be perfectly dry when gathered, and the box should be well filled to prevent shaking and bruising.

PRESERVING THE FRUIT.

For this purpose, the fruit must be thoroughly ripe. When fully ripe, the stem will turn brown, and shrivel somewhat. The fruit is then carefully gathered, and laid upon a dry floor, or shelves, for a day or two, so that some of the moisture will evaporate. They can then be packed in boxes, in about the same manner as described before, but paper will be better than leaves for this purpose. They are then put away on shelves, in an airy room, which must, however, be free from frost, in an even temperature of from 30?to 40? They should be examined from time to time, and the decayed berries taken out. They may thus be kept for several months.

GATHERING THE FRUIT TO MAKE WINE.

For this purpose, the grapes should hang as long as it is safe to allow them; for it will make a very material difference in the quality of the wine, as the water will evaporate, and only the sugar remain; and the flavor or the bouquet will only be fully developed in fully ripened fruit. For gathering, use clean tin or wooden pails; cut the stems as short as possible, and clip or pinch out all unripe or rotten berries, leaving none but fully ripe berries on the bunch. The further process will be described under "wine making."

VARIETIES OF GRAPES.

I would here, again remark, that I consider the question of "what to plant" as chiefly a local one, for which I do not presume to lay down fixed rules; but which every one must, to a certain extent, determine for himself, by visiting vineyards as nearly similar in soil and location to the one he intends to plant, and then closely observing the habits of the varieties after planting. Only thus can we obtain certain results; not by following blindly in the footsteps of so-called authorities, who may live a hundred, or a thousand miles from us, and whose success with certain varieties, on soil entirely different from ours, under different atmospheric influences, can by no means be taken by us as evidence of our success under other circumstances.

CLASS 1.--Varieties most generally used.

CONCORD.

Originated with Mr. E. BULL, of Concord, Mass. This variety seems to be the choice of the majority throughout the country, and however much opinions may differ about its quality, nobody seems to question its hardiness, productiveness, health and value as a market fruit. Here it is of very good quality--and our Eastern brethren have no idea what a really well ripened Missouri grown Concord grape is. It seems to become better the further it is grown West and South; an observation which I think applies with equal force to the Hartford Prolific, Norton's Virginia, Herbemont and others.

Bunch large, heavy shouldered--somewhat compact; berries large, round, black, with blue bloom; buttery, sweet and rich here, when well ripened; with very thin skin and tender pulp. A strong and vigorous grower; with healthy, hardy foliage; free from mildew, and but slightly subject to rot; succeeds well in almost any soil; and is, so far, the most profitable grape we grow. A fine market fruit, and also makes a fine, light red wine, which is generally preferred to the Catawba. Can be easily grown from cuttings.

NORTON'S VIRGINIA, (NORTON'S SEEDLING, VIRGINIA SEEDLING).

Originated by DR. N. NORTON, of Richmond, Virginia. This grape has opened a new era in American grape culture, and every successive year but adds to its reputation. While the wine of the Catawba is often compared to Hock, in the wine of Norton's Virginia, we have one of an entirely different character; and it is a conceded fact that the best red wines of Europe are surpassed by the Norton as an astringent, dark red wine, of great body, fine flavor, and superior medical quality. Vine vigorous and hardy, productive; starting a week later in the Spring than the Catawba, yet coloring a week sooner; and will succeed in almost any soil, although producing the richest wine in warm, southern aspects. Bunches medium, compact; berries small, black, sweet and rich; with dark bluish red juice; only moderately juicy. Healthy in all locations, as far as I know, but I doubt its utility in the East, as I do not think the summers warm and long enough. Seems to attain its greatest perfection in Missouri, but is universally esteemed in the West. Very difficult to propagate, as it will hardly grow from cuttings in open air.

[Illustration: FIG. 24. HERBEMONT.--Berries 1/3 diameter.]

HERBEMONT (HERBEMONT MADEIRA, WARREN).

Origin uncertain. Wherever this noble grape will succeed and fully ripen, it is hard to find a better, for table, as well as for wine. Its home seems to be the South; and I think it will become one of the leading varieties, as soon as the new order of things has been fully established, and free, intelligent labor has taken the place of the drudging, dull toil of the slave. It is particularly fond of warm, southern exposures, with light limestone soil, and it would be useless to plant it on soil retentive of moisture. Bunch long, large shouldered and compact; berry medium, black, with blue bloom--"bags of wine," as Downing fitly calls them; skin thin, sweet flesh, without pulp, juicy and high flavored, never clogs the palate; fine for the table, and makes an excellent wine, which should be pressed immediately after mashing the grapes, when it will be white, and of an exquisite flavor; generally ripens about same time as Catawba. A very vigorous and healthy grower, but tender in rich soils, and should be protected in winter. Extremely productive.

HARTFORD PROLIFIC.

Raised by Mr. STEEL, of Hartford, Conn.: hardy, vigorous and productive; bunch large, shouldered, rather compact; berry full medium, globular, with a perceptible foxy flavor; skin thick, black, covered with blue bloom; flesh sweet, juicy; much better here than at the East; of very fair quality for its time of ripening; hangs well to the bunch here, although said to drop at the East. For market, this is perhaps as profitable as any variety known, as it ripens very early and uniformly, producing immense crops. I have made wine from it, which, although not of very high character, yet ranks as fair.

CLINTON.

Origin uncertain; from Western New York; vigorous, hardy and productive; free from disease; bunch medium, long and narrow, generally shouldered, compact; berry medium, roundish oblong, black, covered with bloom; juicy; somewhat acid; colors early, but should hang late to become thoroughly ripe; brisk vinous flavor, but somewhat of the aroma of the frost grape; makes a dark red wine, of good body, and much resembling claret, but not equal to Norton's Virginia, or even the Concord, in my estimation. Although safe and reliable, I think it has lately been over praised as a wine grape, and as it is a very long, straggling grower, it is one of the hardest vines to keep under control. Propagates with the greatest ease.

DELAWARE.

First disseminated and made known to the public by Mr. A. THOMPSON, of Delaware, Ohio. This is claimed by many to be the best American grape; and although I am inclined to doubt this, and prefer, for my taste, a well ripened Herbemont, it is certainly a very fine fruit. Unfortunately, it is very particular in its choice of soil and location, and it seems as if there are very few locations at the West where it will succeed. Whoever has a location, however, where it will grow vigorously and hold its leaves, will do well to plant it almost exclusively, as it makes a wine of very high character, and is very productive. A light, warm soil seems to be the first requisite, and the bluffs on the north side of the Missouri river seem to be peculiarly adapted to it, while it will not flourish on those on the south side. Bunch small, compact, and generally shouldered; berry below medium, round; skin thin, of a beautiful flesh-color, covered with a lilac bloom; very translucent; pulp sweet and tender, vinous and delicious; wood very firm; short-jointed; somewhat difficult to propagate, though not so much so as Norton's Virginia. Subject in many locations, to leaf-blight, and is there a very slow grower. Fine for the table, and makes an excellent white wine, equal to, if not superior, to the best Rhenish wines, which sells readily at from five to six dollars per gallon. Although I cannot recommend it for general cultivation, it should be tried every where, and planted extensively where it will succeed. Ripens about five days later than Hartford Prolific.

CLASS 2.--Healthy varieties promising well.

CYNTHIANA (RED RIVER).

Origin unknown--said to come from Arkansas. This grape promises fair to become a dangerous rival to Norton's Virginia, which variety it resembles so closely in wood and foliage, that it is difficult if not impossible to distinguish it from that variety. The bunch and berry are of the same color as Norton's Virginia, but somewhat larger, and more juicy; sweeter, with not quite as much astringency, and perhaps a few days earlier. Makes an excellent dark red wine, with not as much astringency, but even more delicate aroma, and was pronounced the "best red wine on exhibition," at the last meeting of the State Horticultural Society, where it was in competition with eight samples of the Norton's Virginia. A strong grower, and productive; as difficult to propagate as the Norton. Mr. FULLER evidently has not the true variety, when he calls it worthless, and identical with the Chippewa and Missouri, from both of which it is entirely distinct.

ARKANSAS.

Closely resembles the foregoing, and will also make an excellent wine of a similar character. I consider both of these varieties as great acquisitions, as they are perfectly healthy, very productive, and will make a wine unsurpassed in merit by any of their class.

TAYLOR (BULLITT.)

This grape, under proper treatment, has proved very productive with me, and will make a wine of very high quality. The bunches and berries are small, it is true; but not much more so than the Delaware; it also sets its fruit well, and as it is hardy, healthy, and a strong grower, it promises to be one of our leading wine grapes. Bunches small, but compact, shouldered; berry small; white at the East; pale flesh-color here; round, sweet, and without pulp; skin very thin. Requires long pruning on spurs, to bring out its fruitfulness.

[Illustration: FIG. 25. HARTFORD PROLIFIC.--Berries 1/2 diameter.]

MARTHA.

This new grape, grown from the seed of the Concord, by that enthusiastic and warm-hearted horticulturist, SAMUEL MILLER, of Lebanon, Pa., promises to be one of the greatest acquisitions to our list of really hardy and good grapes, which have lately come before the public. It has fruited with me the last extremely unfavorable season, and has stood the hardest test any grape could be put to, without flinching. Bunch medium, but compact and heavy, shouldered; berry pale yellow, covered with a white bloom; perhaps a trifle smaller than the Concord; round; pulpy, but sweet as honey, with only enough of the foxy aroma to give it character; juicy--very good. I esteem it more highly than any other white grape I have, as it has the healthy habit and vigorous growth of its parent, and promises to make an excellent white wine. Hangs to the bunch well, and will ripen some days before the Concord.

MAXATAWNEY.

Another very promising white grape--a strong grower, and healthy; may be somewhat too late in the east, but will, I think, be valuable at the West and South. Bunch medium to large---not shouldered; berry above medium; oval; pale yellow, with a slight amber tint on one side; pulp tender, sweet and sprightly; few seeds; fine aroma; quality, best. Ripens about same time as Catawba; seems to be productive.

ROGERS' HYBRID, NO. 1.

This variety, which is also too late in ripening for the East, to be much esteemed there, fruited with me last season, and more than fulfilled all the expectations I entertained of it. It is the best of Mr. ROGERS' Hybrids, which I have yet tasted; and its productiveness, healthy habit, large berry, and good quality, makes it one of the most desirable of all the grapes we raise here, for the table and market. Bunch medium, loose, shouldered; berry very large, oblong, pale flesh-color; skin thin; pulp tender; few seeds, separating freely from the pulp; sweet, vinous and juicy; quality very good. Ripens about same time as Catawba. It is to be regretted that Mr. ROGERS has not named some of the best of his hybrids, as the numbers give rise to many mistakes, and a great deal of confusion. It would be in the interest of grape-growing if this was avoided, by naming at least the best of them.

CREVELING, (CATAWISSA) (BLOOM).

This grape, although not quite perhaps so early as has been claimed for it--ripening about five days after Hartford Prolific--is yet of much better quality; and if it only should prove productive enough, will no doubt make an excellent wine. Bunch long, loose, shouldered; berry full medium, black, round, with little bloom; pulp tender; dark juice, sweet and very good--seems to be hardy and healthy.

NORTH CAROLINA SEEDLING.

Bunch large, shouldered, compact; berry large, oblong, black, with blue bloom; pulpy, but sweet and good; ripens only a few days after Hartford Prolific--very productive, hardy and healthy; strong grower. One of the most showy market grapes we have--not much smaller than Union Village--and as it ripens evenly, and is of very fair quality, is quite a favorite in the market. Makes also a wine of very fair quality.

CUNNINGHAM.

For the West, and very likely further South, this is a very desirable grape for wine, of the Herbemont class. Bunch compact and heavy, sometimes shouldered; berry rather small, black, without pulp, juicy sweet and good;

productive, but somewhat tender; strong grower; should be covered in Winter; makes a very delicious wine, of the Madeira class, which very often remains sweet for a whole year. Ripens late, about a week after the Catawba.

RULANDER.

Mr. FULLER evidently does not know this grape, as he says it is the same as Logan. The Rulander we have here, is claimed to be a true foreign variety. I am inclined to think, however, that it is either a seedling from foreign seed, raised in the country, or one of the Southern grapes of the Herbemont class. Be this as it may however, it certainly bears no resemblance to the Logan, which is a true Fox, of the Labrusca family. Vine a strong, vigorous, short-jointed grower, with heart shaped, light green, smooth leaves; very healthy, and more hardy than either the Herbemont or Cunningham. Bunch rather small, very compact, shouldered; berry small, black, without pulp, juicy sweet and delicious; not subject to rot or mildew: makes a delicious, high flavored wine, but not a great deal of it. The wine of this variety is certainly one of the most delicate and valuable ones we have yet made here and on the soil around Hermann, it will, I think, take preference over the Delaware. Ripens a few days later than Concord.

LOUISIANA (BURGUNDER).

Introduced here by Mr. F. MUENCH, who received it from Mr. THEARD, of Louisiana, where it has been cultivated for some time. Some claim that it is the grape which makes the famous white Burgundy wine of Europe. I am inclined to think it is also a native, grown from foreign seed, like the foregoing, which it closely resembles in foliage and wood; but will, I think, make a wine of still higher quality, perhaps the most delicate white wine we yet have. It can hardly be distinguished from the Rulander in appearance, but has a more sprightly flavor. Ripens at the same time.

ALVEY (HAGAR).

This nice little grape will certainly make one of the most delicious red wines we have, if it can only be raised in sufficient quantity. It is healthy and moderately productive, but a slow grower. Bunch loose, small, shouldered; berry small, black, without pulp, juicy, sweet and delicious; quality best.

Ripens about the same time as the Concord.

CASSADY.

Bunch medium, very compact, shouldered; berry medium, round, greenish-white, covered with white bloom; thick skin, pulpy, but very sweet, and of fine flavor; makes an excellent white wine; very productive, but somewhat subject to leaf-blight in wet seasons; does not rot or mildew.

[Illustration: FIG. 26. CONCORD.--Berries 1/2 diameter.]

BLOOD'S BLACK.

Has often been confounded with Mary Ann, as both varieties were disseminated here, by different persons, under the same name. The true Blood's Black is a few days later than Hartford Prolific; bunch heavy and compact, shouldered; berry round, black, full medium, of very fair quality, and an excellent early market grape. The vine is healthy, hardy, and enormously productive.

UNION VILLAGE.

Perhaps the largest native grape, of fair quality; bunch large, heavy and compact, shouldered; berry very large, oval, black, with blue bloom, pulpy, but juicy, sweet and good. Of better quality here than Isabella; tolerably free from disease, and a splendid market and table fruit. Ripens rather late.

PERKINS.

For those who do not object to a good deal of foxy flavor, this will be a valuable market grape, on account of its earliness, beautiful color, and great productiveness. Mr. FULLER has evidently not the true variety, as he describes it as a "black grape, sour and worthless."

Bunch medium, compact, shouldered; berry full medium, oval, flesh-color, with a beautiful lilac bloom; very sweet, pulpy and foxy. Ripens at same time with Hartford Prolific. Vine a strong grower, healthy and hardy.

CLARA.

For family use, there is at present no grape here at the West, which is superior to this in quality; and although it will not pay to plant largely, either for market or wine, yet no one who can appreciate a really good grape, should be without a few vines of it at least.

Bunch long, rather loose, shouldered; berry medium, pale yellow, translucent, without pulp, sweet, juicy, and of excellent flavor; vine moderately productive and healthy. Ripens with Catawba.

IVES' SEEDLING, (IVES' MADEIRA).

This variety is recommended so much lately, as a superior grape for red wine, that I will mention it here, although I have not yet fruited it. It was first introduced by Col. WARING, of Hamilton County, Ohio, and is said to be free from rot, healthy and vigorous, and to make an excellent red wine, the must having sold from the press at $4 to $5 per gallon. The following description is from bunches sent me from Ohio last fall:

Bunch medium, compact, shouldered; berry rather below medium, black, oblong, juicy, sweet and well flavored; ripens about the time of the Concord. Vine vigorous and healthy; said to propagate with the greatest ease; evidently belonging to the Labrusca species.

We have a seedling here of the Norton's Virginia, raised by Mr. F. LANGENDORFER, of this neighborhood, which promises to be a valuable wine grape for this location. It has not yet been named, and the owner says will never receive a name, unless it proves, in some respect, superior to anything we have yet. He has fruited it twice, and made wine from it the last season, which is of a very high character, resembling Madeira, of a brownish-yellow color; splendid flavor, and of great body. The vine is a strong grower, healthy and very productive; bunch long, seldom shouldered, very compact; berry small, black, with blue bloom; only moderately juicy, and ripens a week later than its parent. I am inclined to think that it will be of great value here and further south as a wine grape, although it would ripen too late to suit the climate further north.

It may be expected here that I should speak of the Iona, Israella, and Adirondac, as many, and good authorities too, think they will be very valuable. The Iona and Israella have fruited but once with me, last summer, and my experience, therefore, has not been long enough to warrant a decided opinion. As far as it goes, however, it has been decidedly unfavorable. My Iona vine set about twenty five bunches, but mildewed and rotted so badly, that I hardly saved as many berries. It may improve in time, but I hardly think it will do for our soil; whatever it may do for others--and I cannot put it down as "promising well." It is a grape of fine quality, where it will succeed. The Israella stood the climate and bad weather bravely, but ripened at least five days later than the Hartford Prolific close by, and was not as good in quality as that grape; in fact, the most insipid and tasteless grape I ever tried. They may both improve, however, upon closer acquaintance, or be better in other locations. Here, I do not feel warranted in praising them, and a description will hardly be needed, as their originator has taken good care to so fully bring their merits, real or imaginary, before the grape-growing community, that it would be superfluous for me to describe them.

The Adirondac I saw and much admired at the East, in 1863; and if its originator, Mr. BAILEY, had only been liberal enough to furnish me with a scion of two eyes, for which I offered to pay him at the rate of a dollar per eye, I would, perhaps, be able to report about it. Instead of the scion, he sent me a dried up vine, which had no life in it when I received it, and in consequence of these disadvantages, I have not been able to fruit it yet. It seems to be healthy and vigorous, however; and should the quality of the fruit be the same as at the East, may be a valuable acquisition.

On this list I have only mentioned those which have fruited here from four to five years, with very few exceptions, and which have generally, during that time, proved successful. To fully warrant the recommendation of a grape for general cultivation I think, we should have fruited it at least five or six years; and although there are many on this list which I should not hesitate to plant largely, yet I have preferred to be rather a little over cautious than too sanguine.

CLASS 3.--Healthy varieties, but inferior in quality.

MINOR SEEDLING, (VENANGO).

This grape has attracted some attention lately--some persons claiming for it superior qualities as a wine grape, even classing it with the Delaware, a statement which I cannot believe. It is a rank Fox, and I can therefore hardly think it will make a wine to suit a fastidious palate.

Bunch medium, very compact, sometimes shouldered; berry full medium, pale red, round, sweet, but very pulpy and foxy. Ripens later than Catawba; is very productive, vigorous and healthy--not subject to rot.

MARY ANN.

The earliest grape we have--healthy, hardy and productive--but in point of quality, a rather poor Isabella, which it much resembles.

Bunch full medium, moderately compact, shouldered; berry medium, oval, black, pulpy, with a good deal of acidity, and strong flavor. Ripens about four to five days before the Hartford Prolific, but is much inferior to that variety in quality.

NORTHERN MUSCADINE.

Very productive and healthy, but too foxy, and liable to drop from the bunch when ripe.

Bunch medium, compact, sometimes shouldered; berry round, brown, sweet, very foxy--pulpy. Ripens about five days later than Hartford Prolific.

LOGAN.

Ripens about same time with Hartford Prolific--but rather inferior in quality. Bunch long, loose, shouldered; berry medium, oval; resembling Isabella.

BROWN.

Resembling Isabella, but more free from disease; good grower and productive; will suit those who like the Isabella.

HYDE'S ELIZA, (CANBY'S AUGUST).

Bunch medium, compact; berry medium, round, black, juicy; rather pleasant, but unproductive, and of little value, where better varieties can be had.

MARION PORT.

Resembles the foregoing; may, perhaps, make a better wine, but cannot be recommended.

POESCHEL'S MAMMOTH.

Grown here, from seed of the Mammoth Catawba, by Mr. MICHAEL POESCHEL.

Bunch medium, compact, sometimes shouldered; berry very large, round, pale red, pulpy; rather deficient in flavor, but very large; free from disease. Ripens a week later than Catawba.

CAPE (ALEXANDER, SCHUYLKILL MUSCADELL).

Bunch rather small, compact; berry medium, black, round, pulpy, rather sweet, dark juice. Said to make a good red wine, but my experience has not been favorable. Ripens late--a week after the Catawba.

DRACUT AMBER.

A Fox Grape, pale red, pulpy, inferior in quality and color to Perkins, which it closely resembles; ripens about same time.

ELSINBURGH, (MISSOURI BIRD'S EYE).

This old variety was largely disseminated under the latter name, by NICHOLAS LONGWORTH, of Cincinnati. It is a nice little grape; but too unproductive to be of any value here, although it makes a very superior wine. Bunch long and loose, shouldered; berry small, round, black, moderately juicy, with little pulp, sweet and good. Ripens a week before the Catawba.

GARBER'S ALBINO.

A grape of very fair quality, and rather early, but a shy bearer. Bunch small, rather loose; berry medium, pale yellow, sweet and good.

FRANKLIN.

A strong grower; said to be very productive; resembling Clinton in foliage and general habit. Bunch small, compact; berry below medium, black, juicy, with a marked frost grape flavor, and hardly worthy of cultivation.

LENOIR.

Of the Herbemont class, but about a week earlier; of good quality, but too unproductive to be recommended. Bunch medium, compact, shouldered; berry small, round, black, sweet and good.

NORTH AMERICA.

Early and hardy, but too unproductive, and bunch too small. Bunch small, shouldered; berry round; of very good quality for its season; black, juicy. Ripens as early as Hartford Prolific.

CLASS 4.--Varieties of good quality, but subject to disease.

CATAWBA.

This well known grape was brought into notice by Major ADLUM, of Georgetown, D.C., who thought he had, by its introduction, conferred a greater boon upon the American people, than if he had paid the national debt. For the last ten years, it has been so much subject to disease, that it cannot be recommended any longer, except for some peculiar locations. It is said to be healthy in northern Illinois and Iowa, where it will not stand the winter, however, without protection.

Bunch large, moderately compact, shouldered; berry medium, red, covered with lilac bloom; juicy, pulpy, sweet, somewhat astringent, of good flavor. A fair grape for the table, and makes a good wine, resembling Hock, but subject to mildew, rot and leaf-blight.

DIANA.

A seedling of the foregoing, raised by Mrs. DIANA CREHORE. Perhaps one of the most variable of all the grapes, being very fine one season, and very indifferent the next. Bunch large and long, compact, shouldered; berry pale red, round, somewhat pulpy; thick skin; juicy and sweet, with a peculiar flavor, which DR. WARDER very aptly calls "feline;" others call it "delicate." Very productive, but subject to leaf-blight, mildew and rot; although perhaps not so much as the Catawba. Ripens about a week earlier.

ISABELLA.

Unworthy of cultivation here, but said to be better at the North. Bunch long, loose, shouldered; berry medium, oval, black; tough pulp, with a good deal of acidity, juicy, and a peculiar flavor. Ripens irregularly. Subject to rot and leaf-blight.

GARRIGUES.

Closely resembling the Isabella, but ripens more evenly, and is of somewhat better quality.

TOKALON.

Bunch large, loose, shouldered; berry black, large, sweet and buttery; of very good quality, but very much subject to disease. Ripens somewhat later than Catawba.

ANNA.

Bunch large and loose; berry pale amber, covered with white bloom; sweet, tolerable flavor, but poor bearer, and subject to mildew. Ripens about same time as Catawba.

ALLEN'S HYBRID, (ALLEN'S WHITE HYBRID).

Bunch large and loose, shouldered; berry medium, nearly round; white, without pulp, juicy and delicious; quality very good, but variable; sometimes best. Said to be a hybrid of Vitis Labrusca and a foreign grape, raised by J. F. ALLEN, Salem, Massachusetts, and is really a fine grape, although too tender and variable for extensive vineyard culture. Ripens about two weeks before Catawba.

CUYAHOGA (COLEMAN'S WHITE).

Much recommended in Ohio, where it originated, but unworthy of culture here, being a poor grower, a shy bearer and very much subject to leaf-blight. Bunch medium, compact; berry dirty greenish-white; thick skin; pulpy, and insipid.

DEVEREAUX.

This is, in dry seasons, a really fine grape, but subject to leaf-blight and mildew in hot seasons. Bunch often a foot long, loose, shouldered; berry below medium, round, black, juicy; without pulp, sweet and vinous. Belonging to the Herbemont family; is a strong grower; very productive, and rather tender. May be valuable in well drained soils, and southern climate, as it undoubtedly will make a fine wine.

KINGSESSING.

Bunch long and loose, large, shouldered; berry medium, round, pale red, with fine lilac bloom; pulpy; of fair quality, but subject to leaf-blight, and mildew.

ROGERS' HYBRID, NO. 15.

Bunch large, loose, shouldered; berry above medium, red with blue bloom, roundish-oblong, pulpy, with peculiar flavor, sweet and juicy. A showy grape, but not very good in quality, and much subject to mildew and rot. Ripens at the same time with Catawba.

CLASS 5.--Varieties unworthy of cultivation.

OPORTO.

Of all the humbugs ever perpetrated upon the grape-growing public, this is one of the most glaring. The vine, although a rank and healthy grower, is unproductive; seldom setting more than half a dozen berries on a bunch, and these are so sour, have such a hard pulp, with such a decided frost-grape taste and flavor, and are so deficient in juice, that no sensible man should think of making them into wine, much less call it, as its disseminator did, "the true port wine grape."

MASSACHUSETTS WHITE.

This was sent me some eight years ago, by B. M. WATSON, as "the best and hardiest white grape in cultivation," and he charged me the moderate sum of $5 each, for small pot plants, with hardly two eyes of ripened wood. After careful nursing of three years, I had the pleasure of seeing my labors rewarded by a moderate crop of the vilest red Fox Grapes it has ever been my ill luck to try.

The foregoing have all been tried by me, and have been characterized and classified as I have found them here. The following are varieties I have not fruited yet, although I have them on trial.

Varieties highly recommended by good authorities: Telegraph, Black Hawk, Rogers' Hybrids, Nos. 3, 4, 6, 9, 12, 13, 19, 22, 33, Hettie, Lydia, Charlotte, Mottled, Pauline, Wilmington, Cotaction and Miles.

There are innumerable other varieties, for which their originators all claim peculiar merits, and some of whom may prove valuable. But all who bring new varieties before the public, should consider that we have already names enough, nay, more than are good for us, and that it is useless to swell the list still more, unless we can do so with a variety, superior in some respects to our best varieties. A new grape, to claim favor at the hands of the public, should be healthy, hardy, a good grower, and productive; and of superior quality, either for the table or for wine.

There are some varieties circulated throughout the country as natives, which are really nothing but foreign varieties, or, perhaps, raised from foreign seed. They will not succeed in open air, although now and then they will ripen a bunch. The Brinkle, Canadian Chief, Child's Superb, and El Paso belong to this class.

A really good table grape should have a large amount of sugar, but tempered and made more agreeable by a due proportion of acid, as, if the acid is wanting, it will taste insipid; a tender pulp, agreeable flavor, a large amount of juice, a good sized bunch, large berry, small seeds, thin skin, and hang well to the bunch.

A good wine grape should have a large amount of sugar, with the acid in due proportion, a distinctive flavor or aroma; though not so strong as to become disagreeable, and for red wines a certain amount of astringency. It is an old vintner's rule, that the varieties with small berries will generally make the best wine, as they are generally richer in sugar, and have more character than varieties with larger berries.

WINE-MAKING.

GATHERING THE GRAPES.

Although I have described the process already, I will here again reiterate that the grapes should be thoroughly ripe. This does not simply mean that they are well colored. The Concord generally begins to color here the 5th of August, and we could gather the majority of our grapes, of that variety, for market, by the 15th or 20th of that month; but for wine-making we allow them to hang until the 15th or 20th of September, and sometimes into October. Thus only do we get the full amount of sugar and delicacy of aroma which that grape is capable of developing, as the water evaporates, and the sugar remains; it also loses nearly all the acidity from its pulp; and the latter, which is so tough and hard immediately after coloring, nearly all dissolves and becomes tender. The best evidences of a grape being thoroughly ripe are: 1st. The stem turns brown, and begins to shrivel; 2nd, the berry begins to shrivel around the stem; 3d, thin and transparent skin; 4th, the juice becomes very sweet, and sticks to the finger like honey or molasses, after handling the grapes for some time.

It is often the case that some bunches ripen much later on the vines. In such a case, the ripest should be gathered first, and those that are not fully ripe remain on the vines until mature. They will ripen much quicker if the ripest bunches have been removed first.

The first implements needed for the gathering are clean wooden and tin pails and sharp knives, or better still, the small shears spoken of in a former part of this work. Each gatherer is provided with a pail, or two may go together, having a pail each, so that one can empty and the other keep filling during the time. If there are a good many unripe berries on the bunches, they may be put into a separate pail, and all that are soft will give an inferior wine. The bunch is cut with as short a stem as possible, as the stem contains a great deal of acid and astringency; every unripe or decayed berry is picked out, so that nothing but perfectly sound, ripe berries remain.

The next implement that we need is a wooden tub or vat, to carry the grapes to the mill; or the wagon, if the vineyard is any distance from the cellar. This is made of thin boards, half-inch pine lumber generally; 3 feet high inside, 10 inches wide at the bottom, 20 inches wide at the top, being flat on one side, where it is carried on the back, and bound with thin iron hoops. It is carried by two leather-straps running over the shoulders, as shown in Fig. 29, and should contain about eight to ten pails, or a little over two bushels of grapes. The carrier can pass easily through the rows with it to any part of the vineyard, and lean it against a post until full. If the vineyard is close to the cellar or press-house, the grapes can be carried to it directly; if too far, we must provide a long tub or vat, to place on the wagon, into which the grapes are emptied. I will here again repeat that the utmost cleanliness should be observed in all the apparatus; and no tub or vat should be used that is in the least degree mouldy. Everything should be perfectly sweet and clean, and a strict supervision kept up, that the laborers do not drop any crumbs of bread, &c., among the grapes, as this will immediately cause acetous fermentation. The weather should be dry and fair, and the grapes dry when gathered.

THE WINE-CELLAR.

As the wine-cellar and press-house are generally built together, I will also describe them together. A good cellar should keep about an even temperature in cold and warm weather, and should, therefore, be built sufficiently deep, arched over with stone, well ventilated, and kept dry.

Where the ground is hilly, a northern or northwestern slope should be chosen, as it is a great convenience, if the entrance can be made even with the ground. Its size depends, of course, upon the quantity of wine to be stored. I will here give the dimensions of one I am constructing at present, and which is calculated to store from 15,000 to 20,000 gallons of wine. The principal cellar will be 100 feet long, by 18-1/2 feet wide inside, and 12 feet high under the middle of the arch. This will be divided into two compartments; the back one, at the farthest end of the cellar, to be 40 feet, which is destined to keep old wine of former vintages; as it is the deepest below the ground, it will keep the coolest temperature. It is divided from the front compartment by a wall and doors, so that it can be shut off should it become necessary to heat the other, while the must is fermenting. The other compartment will be 60 feet long, and is intended for the new wine, as the temperature will be somewhat higher, and, therefore, better adapted to the fermentation of the must. This will be provided with a stove, so that the air can be warmed, if necessary, during fermentation. This will also be closed by folding doors, 5-1/2 feet wide. There will be about six ventilators, or air-flues, on each side of these two cellars, built in the wall, constructed somewhat like chimneys, commencing at the bottom, whose upper terminus is about two feet above the arch, and closed with a grate and trap-doors, so that they can be closed and opened at will, to admit air and light. Before this principal cellar is an arched entrance, twenty feet long inside, also closed by folding doors, and as wide as the principal cellar. This will be very convenient to store empty casks, and can also be used as a fermenting room in Fall, should it be needed. The arch of the principal cellar will be covered with about six feet of earth; the walls of the cellar to be two feet thick. The press-house will be built above the cellar, over its entire length, and will also be divided into two rooms. The part farthest from the entrance of the cellar, to be 60 feet by 18, will be the press-house proper, with folding doors on both sides, about the middle of the building, and even with the surface ground, so that a wagon can pass in on one side and out on the other. This will contain the grape-mill, wine-presses, apparatus for stemming, and fermenting vats for white or light-colored wine. The other part, 40 feet long, will contain an apparatus for distilling, the casks and vats to store the husks for distilling, and the vats to ferment very dark colored wines on the husks, should it be necessary. It will also be used as a shop, contain a stove, and be floored, so that it will be convenient, in wet and cold weather, to cut cuttings, &c. A large cistern, to be built on one side of the building, so that the necessary water for cleaning casks, &c., will be handy;

with a force-pump, will complete the arrangement. I need hardly add here, that the whole cellar should be paved with flags or brick, and well drained, so that it will be perfectly dry.

This cellar is destined to hold two rows of casks, five feet long, on each side. For this purpose layers of strong beams are provided, upon which the casks are laid in such a manner that they are about two feet from the ground, fronting to the middle, and at least a foot or eighteen inches of space allowed between them and the wall, so that a man can conveniently pass and examine them. This will leave five and a-half to six feet of space between the two rows, to draw off the wine, move casks, &c.

This cellar will, at the present rates of work, cost about $6,000. Of course, the cellar, as before remarked, can be built according to the wants of the grape-grower. For merely keeping wine during the first winter, a common house cellar will do; but during the hot days of summer wine will not keep well in it.

APPARATUS FOR WINE-MAKING.--THE GRAPE MILL AND PRESS.

This mill can be made very simple, of two wooden rollers, fastened in a square frame, running against each other, and turned with a crank and cog-wheel. The rollers should be about nine inches in diameter, and set far enough apart to mash the berries, but not the seeds and stems. A very convenient apparatus, mill and press, is manufactured by Geiss & Brosius, Belleville, Ill., and where the quantity to be made does not exceed 2,000 gallons, it will answer every purpose. The mill has stone rollers, which can be set by screws to the proper distance, with a cutting apparatus on top, for apples in making cider, which can be taken off at will. The press is by itself, and consists of an iron screw, coming up through the platform, with a zinc tube around it to prevent the must from coming in contact with it. The platform has a double bottom, the lower one with grooves; the upper consists simply of boards, with grooves through it to allow the must to run through. These boards are held in their places by wooden pegs, and can be taken off at will. A circular hopper, about a foot in diameter, and made of laths screwed to iron rings, with about a quarter of an inch space between them, encloses the zinc tube. The outer frame is constructed in the same way, is about 2-1/2 feet in diameter, and bound with strong wooden and iron hoops. The mashed grapes are poured into the frame, a close-fitting cover is put on, which is held down by a strong block, and the power is applied by an iron nut just on the top of the screw, with holes in each end to apply strong wooden levers. The apparatus is strong, simple, and convenient, and presses remarkably fast and clean, as the must can run off below, on the outside and also on the inside. The cost of mill and press is about $90, but each can be had separately for $45.

If a large amount of grapes are to be pressed, the press should be of much larger dimensions, but may be constructed on the same principle--a strong, large platform, with a strong screw coming through the middle, and a frame made of laths, screwed to a strong wooden frame, through which the must can run off freely, with another frame around the outside of the platform. The must runs off through grooves to the lower side, where it is let off by a spout. It may be large enough to contain a hundred bushels of grapes at a single pressing, for a great deal depends upon the ability of the vintner to press a large amount just at the proper time, when the must has fermented on the husks just as long as he desires it to do.

FERMENTING VATS.

These should correspond somewhat with the size of the casks we intend to fill; but they are somewhat unhandy if they hold more than, say four hundred gallons. They are made of oak or white pine boards, 1-1/2 inch thick, bound securely by iron hoops, about three feet high, and, say, five feet wide. The bottom and inside must be worked clean and smooth, to facilitate washing. When the must is to ferment a longer time on the husks, as is often the case in red wines, a false bottom should be provided, for the purpose of holding the husks down below the surface of the must. It is made to fit the size of the vat, and perforated with holes, and held in its place by sticks of two inches square, let into the bottom of the vat, and which go through the false bottom. A hole is bored through them, and the bottom held down by means of a peg passed through this hole. The vat is closed by a tight-fitting cover, through which a hole is bored, large enough to admit a tin tube of about an inch in diameter, to let off the gas. The vats are set high enough above the ground to admit drawing off the must through a faucet near the bottom of the vat. For those grapes which are to be pressed immediately we need no false bottoms or covers for the vats. As fermentation generally progresses very rapidly here, and it is not desirable with most of our wines to ferment them on the husks very long, as they generally have astringency enough, operations here are much more simple than in Europe.

The must is generally allowed to run into a large funnel, filled with oat straw, and passes through a hose into the casks in the cellar. A hole can be left through the arch for that purpose, as it is much more convenient than to carry the must in buckets from the press into the casks.

It is sometimes desirable to stem the grapes, although it is seldom practiced in this country. This can be easily done by passing the bunches rapidly over a grooved board, made somewhat in the form of a common washboard, only the grooves should be round at the bottom and the edges on top. It is seldom desirable here.

THE WINE CASKS.

These should be made of well-seasoned white oak staves, and can, of course, be of various sizes to meet the wants of the vintner. The best and most convenient size for cellar use I have found to be about 500 gallons. These are sufficiently large to develop the wine fully, and yet can be filled quick enough to not interrupt fermentation. Of course, the vintner must have some of all sizes, even down to the five-gallon keg; but for keeping wine, a cask of 500 gallons takes less room comparatively, and the wine will attain a higher degree of perfection than in smaller casks. The staves to make such a cask should be about 5 feet long, and 1-1/2 to 2 inches thick, and be the very best wood to be had. The cask will, when ready, be about as high as it is long, should be carefully worked and planed inside, to facilitate washing and have a so-called door on one end, 12 inches wide and 18 inches high, which is fastened by means of an iron bolt and screw, and a strong bar of wood. This is to facilitate cleaning; when a cask is empty, the door is taken out, and a man slips into the cask with a broom and brush, and carefully washes off all remnants of lees, etc., which, as the lees of the wine are very slimy and tenacious, cannot be removed by merely pouring in water and shaking it about. It is also much more convenient to let these large casks remain in their places, than to move them about. The casks are bound with strong iron hoops.

To prepare the new casks, and also the vats, etc., for the reception of the must, they should be either filled with pure water, and allowed to soak for several days, to draw out the tannin; then emptied, scalded with hot water, and afterwards steamed with, say two or three gallons of boiling wine; or they can be made "wine-green," by putting in about half a bushel of unslaked lime, and pouring in about the same quantity of hot water. After the lime has fallen apart, add about two quarts of water to each pound of lime, put in the bung, and turn the cask about; leaving it lie sometimes on one side, sometimes on the other, so that the lime will come in contact with every part of the cask. Then pour out the lime-water; wash once or twice with warm water, and rinse with a decoction of vine leaves, or with warm wine. Then rinse once more with cold water, and it will be fully prepared to receive the must. This is also to be observed with old casks, which have become, by neglect or otherwise, mouldy, or have a peculiar tang.

MAKING THE WINE.

As we have our apparatus all prepared now, we can commence the operation itself. This can be done in different ways, according to the class of wine we are about to make.

To make white, or light-colored wine, the grapes which were gathered and mashed during the day, can be pressed and put into the cask the following night. To mash them, we place the mill above one of the fermenting vats, mashing them as quick as they are carried or hauled to the press-house. The vat is simply covered with a cloth during the day. If the season has been good, the must will make good wine without the addition of anything else. In poor seasons it will be necessary to add water and sugar, to improve its quality, but I will speak of this method in a separate chapter. In the evening, the must which will run off, is first drawn from the vat, and by some kept separate; but I think, it makes, upon the whole, a better wine, if the pressing is added to it. The husks, or mashed grapes, are then poured upon the press, and pressed until fully dry. To accomplish this the press is opened several times, and the edges of the cake, or "cheese," as some call it, are cut off with an axe or cleaver and put on top, after which they are pressed down again. The casks are then filled with the must; either completely, if it is intended that the must should ferment above, as it is called, or under, when the cask is not completely filled, so that the husks, which the must will throw up, will remain in the cask. Both methods have their advantages, but I prefer the former, with a very simple contrivance, to exclude the air, and also prevent waste. This is a siphon or tin tube, bent in the form of a double elbow, of which one end fits tightly in the bung hole, and the other empties into a dish of water, to be set on one end of the cask, through which the gas escapes, as shown in Fig. 30.

We should, however in pressing, be guided somewhat by the weather. In warm weather fermentation will commence much sooner, and be more violent, than when the weather is cold. Consequently we should press much sooner in warm weather, than when the air is cool. Late in the fall, it is sometimes advisable to leave the must a day longer on the husks, than indicated below. The cellar should be kept at an even temperature of about 60?during the first few weeks, and if it does not naturally attain this temperature, then it should be warmed by a stove, as much of the quality of the wine depends upon a thorough fermentation during the first ten days.

When violent fermentation has ceased, say after about ten or twelve days, and the must has become quiet, the cask should be closed with a tight bung, and the wine is left until it is clear. In about two to three months it ought to be perfectly clear and fine--is then racked, i.e., drawn from the lees, by means of a faucet, and put into clean, sweet casks. It is very important that the casks are "wine-seasoned," that is, have no other tang than of wine. For must, fresh brandy or whiskey casks may be used, but after the wine has fermented, it will not do to use such, as the wine will acquire the smell and taste of the liquor. When a cask has been emptied, it should be carefully cleaned, as before described, by entering at the door, or with smaller casks, by taking out the head. After it is thoroughly cleansed, it may be fumigated slightly, by burning a small piece of sulphured paper, or a nutmeg in it, and then filled. To keep empty casks in good condition they should, after cleaning, be allowed to become thoroughly dry, when they are sulphured, closed tightly, and laid away in the cellar. The operation of sulphuring should be repeated every six weeks. If wanted for use, they are simply rinsed with cold water.

For racking the wine, we should have: 1st a large brass faucet. 2d. Pails of a peculiar shape, wider at the top, to prevent wastage. 3d. A wooden funnel, as shown in Fig. 31, to hold about six gallons. In racking--first carefully lift the bung of the cask, as the exclusion of air from above would cause a gurgling motion in the cask, if tapped below, which would stir up the lees in the bottom. Then, after having loosened with a hammer the wooden peg, closing the tap hole, let your assistant hold the pail opposite the hole, hold the faucet in your right hand, and with the left, withdraw the plug, inserting the faucet quickly. Drive it in firmly with a hammer, and you are ready for the work.

Do not fully open the faucet at first, because the first pailful is generally not quite clear, and should run slowly. You can keep this by itself; and this, and the last from the lees, is generally put into a cask together and allowed to settle again. It will make a good, clear wine after a few weeks. As soon as the wine runs quite clear and limpid, it can be put into the cask destined to receive it, and you can let it run as fast as it can be emptied. When the wine has run off down to the tap hole, the cask may be carefully raised on the other end, one inserting a brick or piece of board under it, while the other lifts gently and slowly. This may be repeated several times, as long as the wine runs clear; and even the somewhat cloudy wine may be put with the first pailful into a separate cask. As soon as it comes thick or muddy, it is time to stop. The lees are emptied out, and will, if distilled, make a fine flavored and very strong brandy.

This treatment can be applied to all white and light-colored wines, when it is not desirable to have a certain astringency in the wine. The Catawba, Concord, Herbemont, Delaware, Rulander, Cassady, Taylor, Louisiana, Hartford Prolific, and Cunningham should all be treated in a similar manner. The Concord, although it will, under this treatment, make only a light red wine, of which the color can be changed to dark red by fermenting on the husks, is not desirable if treated in the latter manner; as the peculiar foxy aroma of the grape will be imparted to the must to such a degree, as to make the flavor disagreeable, I shall recur to the subject of flavor in wines in another chapter.

To make red wine, the must should be fermented on the husks, as generally the darkest color is desired, and also, a certain astringency, which the wine will acquire principally from the seeds, skins, and stems of the grapes, which contain the tannin. The grapes are mashed, and put into the fermenting vat, of the kind described before, with false bottoms. After the vat is filled about three-fourths the false bottom is put on, the husks are pressed down by it, until they are covered about six inches by the must, and the cover put on. It is seldom desirable here to ferment longer than three days on the husks, if the weather is warm--in a temperature of 60?-two days will often be enough, as the wine will become too rough and astringent by an excessively long fermentation. Only experience will be the proper guide here, and also the individual taste. It will be generally time to press, when the must has changed its sweet taste, and acquired a somewhat rough and bitter one. Where it is desired to make a very dark colored wine, without too much astringency, the grapes should be stemmed, as most of the rough and bitter taste is in the stems; and it can then be fermented on the husks for six or eight days. In this manner the celebrated Burgundy wines are made; also most of the red wines of France and Germany. Many of them are even allowed to go through the whole process of fermentation, and the husks are filled into the cask with the must, through a door, made in the upper side of the cask; and it there remains, until the clear wine is drawn off. This is seldom desirable here, however, as our red wine grapes have sufficient astringency and color without this process. The treatment during fermentation, racking, etc., is precisely the same as with white wine, with only this difference, that the red wine is generally allowed to stay longer on the lees; for our object in making this class of wine is different than in making white, or so-called Schiller or light red wine. In white and light colored wines we desire smoothness and delicacy of bouquet and taste; in dark red wines, we desire astringency and body, as they are to be the so-called stomach or medical wines. It is therefore generally racked but once, in the latter part of February or March, and the white and light colored wines are racked in December or January, as soon as they have become clear--and again in March. We also use no sulphur in fumigating the casks, as it takes away the color to a certain extent. We generally do not use anything, but simply clean the casks well, in racking red wine.

I will say a few words in regard to under fermentation. If this method Is to be followed, the casks are not filled, but enough space left to allow the wine to ferment, without throwing out lees and husks at the bung. The bung is then covered, by laying a sack filled with sand over it, and when fermentation is over--as well by this as by the other method--the casks are filled with must or wine, kept in a separate cask for the purpose. The casks should always be kept well filled, and must be looked over and filled every two or three weeks, as the wine will continually lose in quantity, by evaporation through the wood of the casks. The casks should be varnished or brushed over with linseed oil, as this will prevent evaporation to some extent.

In wine making, and giving the wine its character, we can only be guided by practice and individual taste, as well as the prevailing taste of the consuming public. If the prevailing taste is for light colored, smooth and delicate wines, we can make them so, by pressing immediately, and racking soon, and frequently. If a dark colored, astringent wine is desired, we can ferment on the husks, and leave it on the lees a longer period. There is a medium course, in this as in everything else; and the intelligent vintner will soon find the rules which should guide him, by practice with different varieties.

Among the wines to be treated as dark red, I will name Norton's Virginia, Cynthiana, Arkansas, and Clinton, and, I suppose, Ives' Seedling. It would be insulting to these noble wines to class with them the Oporto, which may make a very dark colored liquid, but no wine worth the name, unless an immense quantity of sugar is added, and enough of water to dilute the peculiar vile aroma of that grape.

AFTER TREATMENT OF THE WINE.

Even if the wine was perfectly fine and clear, when drawn off, it will go through a second fermentation as soon as warm weather sets it--say in May or June. If the wine is clear and fine, however, the fermentation will be less violent, than if it is not so clear, as the lees, which the wine has never entirely deposited; act as they ferment. It is not safe or judicious, therefore, to bottle the wine before this second fermentation is over. As soon as the wine has become perfectly clear and fine again--generally in August or September--it can be bottled. For bottling wine we need: 1st. clean bottles. 2d. good corks, which must first be scalded with hot water, to soften them, and draw out all impurities, and then soaked in cold water. 3d. a small funnel. 4th. a small faucet. 5th. a cork-press, of iron or wood. 6th. a light wooden mallet to drive in the corks.

After the faucet has been inserted in the cask, fill your bottles so that there will be about an inch of room between the cork and the wine. Let them stand about five minutes before you drive in the cork, which should always be of rather full size, and made to fit by compressing it with the press at one end. Then drive in the cork with the mallet, and lay the bottles, either in sand on the cellar floor, or on a rack made for that purpose. They should be laid so that the wine covers the cork, to exclude all air.

The greater bulk of the wine, however, if yet on hand; can be kept in casks. All the wine to be kept thus, should be racked once in about six months, and the casks kept well filled. Most of our native wines, however, are generally sold after the second racking in March, and a great many even as soon as clear--in January.

DISEASES OF THE WINE AND THEIR REMEDIES.

These will seldom occur, if the wine has been properly treated. Cases may arise, however, when it will become necessary to rack the wine, or fine it by artificial means.

TREATMENT OF FLAT AND TURBID WINE.

The cause of this is generally a want of Tannin. If the wine has a peculiar, flat, soft taste, and looks cloudy, this is generally the case. Draw the wine into another cask, which has been well sulphured, and add some pulverized tannin, which can be had in every drug store. The tannin may be dissolved in water--about an ounce to every two hundred gallons of wine--and the wine well stirred, by inserting a stick at the bung. Should it not have become clear after about three weeks, it should be fined. This can be done, by adding about an ounce of powdered gum-arabic to each forty gallons, and stirring the wine well when it has been poured in. Or, take some wine out of the casks--add to each forty gallons which it contains the whites of ten eggs, whipped to foam with the wine taken out--pour in the mixture again--stir up well, and bung up tight. After a week the wine will generally be clear, and should then be drawn off.

USE OF THE HUSKS AND LEES.

These should be distilled, and will make a very strong, fine flavored brandy. The husks are put into empty barrels or vats--stamped down close, and a cover of clay made over them, to exclude the air. They will thus undergo a fermentation, and be ready for distillation in about a month. They should be taken fresh from the press, however; for if they come into contact with the air, they will soon become sour and mouldy. The lees can be distilled immediately. Good fresh lees, from rather astringent wines are also an excellent remedy when the wine becomes flat, as before described.

DR. GALL'S AND PETIOL'S METHOD OF WINE MAKING.

The process of wine making before described, however, can only be applied in such seasons, and with such varieties of grapes, that contain all the necessary elements for a good wine in due proportion. For unfavorable seasons, with such varieties of grapes as are deficient in some of the principal ingredients, we must take a different course--follow a different method. To see our way clearly before us in this, let us first examine which are the constituent parts of must or grape juice. A chemical analysis of must, shows the following result:

Grape juice contains sugar, water, free acids, tannin, gummy and mucous substances, coloring matter, fragrant or flavoring substances, (aroma bouquet). A good wine should contain all these ingredients in due proportion. If there is an excess of one, and a want of the other, the wine will lose in quality. Must, which contains all of these, in due proportion, we call normal must, and only by determining the amount of sugar and acids in this so-called normal must, can we gain the knowledge how to improve such must, which does not contain the necessary proportion of each. The frequent occurrence of unfavorable seasons in Europe, when the grapes did not ripen fully, and were sadly deficient in sugar, set intelligent men to thinking how this defect could be remedied; and a grape crop, which was almost worthless, from its want of sugar, and its excess of acids, could be made to yield at least a fair article, instead of the sour and unsaleable article generally produced in such seasons. Among the foremost who experimented with this object in view I will here name CHAPTAL, PETIOL; but especially DR. LUDWIG GALL, who has at last reduced the whole science of wine-making to such a mathematical certainty, that we stand amazed only, that so simple a process should not have been discovered long ago. It is the old story of the egg of Columbus; but the poor vintners of Germany, and France, and we here, are none the less deeply indebted to those intelligent and persevering men for the incalculable benefits they have conferred upon us. The production of good wine is thus reduced to a mathematical certainty; although we cannot in a bad season, produce as high flavored and delicate wines, as in the best years, we can now always make a fair article, by following the simple rules laid down by DR. GALL. When this method was first introduced, it was calumniated and despised--called adulteration of wine, and even prohibited by the governments of Europe; but, DR. GALL fearlessly challenged his opponents to have his wines analyzed by the most eminent chemists; which was repeatedly done, and the results showed that they contained nothing but such ingredients which pure wine should contain; and since men like VON BABO, DOBEREINER and others have openly endorsed and recommended gallizing, prejudice is giving way before the light of scientific knowledge.

But to determine the amount of sugar and acids contained in the must we need a few necessary implements. These are:

THE MUST SCALE OR SACCHAROMETER.

The most suitable one now in use is the Oechsle's must scale, constructed on the principle that the instrument sinks the deeper into any fluid, the thinner it is, or the less sugar it contains. Fig. 32 shows this instrument, "which is generally made of silver, or German silver, although they are also made of glass. A, represents a hollow cylinder--best made of glass, filled with must to the brim, into which place the must scale B. It is composed of the hollow float a, which keeps it suspended in the fluid; of the weight c, for holding in a perpendicular position; and of the scale e divided by small lines into from fifty to one hundred degrees. Before the gauge is placed in the must, draw it several times through the mouth, to moisten it--but allow no saliva to adhere to it. When the guage ceases to descend, note the degree to which it has sunk; after which press it down with the finger a few degrees further, and on its standing still again, the line to which the must reaches, indicates its so-called weight, expressed by degrees." The must should be weighed in an entirely fresh state, before it shows any sign of fermentation, and should be free from husks, and pure.

This instrument, which is indispensable to every one who intends to make wine, can be obtained in nearly every large town, from the prominent opticians. JACOB BLATTNER, at St. Louis keeps them for sale.

The saccharometer will indicate the amount of sugar in the must, and its use is so simple, that every one can soon become familiar with it. The next step in the improvement of wines was to determine the amount of acids the must contained, and this problem has also been successfully solved by the invention of the acidimeter:

THE ACIDIMETER AND ITS USE.

"The first instrument of this kind which came into general use, was one invented by DR. OTTO, and consists of a glass tube, from ten to twelve inches in length, half an inch in width, and closed at the lower end. Fig. 33 shows OTTO'S Acidimeter.

"The tube is filled to the partition line a, with tincture of litmus. The must to be examined, before it has begun to ferment is then poured into the tube, until it reaches the line 0. The blue tincture of litmus, which would still be blue, if water had been added, is turned into rose-color by the action of the acids contained in the must.

"If a solution of 1,369 per cent, of caustic ammonia is added to this red fluid, and the tube is turned around to effect the necessary mixture, keeping its mouth closed with the thumb, after the addition of more or less of the ammonical fluid, it will change into violet. This tinge indicates the saturation of the acids, and the height of the fluid in the tube now shows the quantity of acid in the must, by whole, half and fourth parts per cent. The lines marked 1, 2, 3, 4, indicate whole per cents.; the short intermediate lines, one-fourth per cents."

When DR. GALL, shortly before the vintage of 1850, first publicly recommended the dilution of the acids, he was obliged to refer to this instrument, as already known, and everywhere at hand, which was at the same time cheap, and simple in its use. "It is true, however, that if must is examined by this instrument, the quantity of acids contained in it, is really somewhat larger than indicated by the instrument; because the acids contained in the must require for their saturation a weaker solution of ammonia than acetic acid." As however, OTTO'S acidimeter shows about one eighth of the acids less than the must actually contains, and about as much acids combined with earths is removed during fermentation, DR. GALL recommends that the quantity of acids be reduced to 6-1/2, or at most 7 thousandths of OTTO'S acidimeter, and the results have shown that this was about the right proportion; as the wines in which the acids were thus diluted were in favor with all consumers.

"The acidimeter referred to was afterwards improved, by making the tube longer and more narrow, and dividing it into tenths of per cents, instead of fourths; thus dividing the whole above 0 into thousandths. But although by this improved acidimeter the quantity of acids could be ascertained with more nicety, there remained one defect, that in often turning the glass tube for mixing the fluids, some of the contents adhered to the thumb in closing its mouth. This defect was remedied in a new acidimeter, invented by Mr. GEISLER, who also invented the new vaporimeter for the determination of the quantity of alcohol contained in wine. It is based on the same principle as OTTO'S, but differs altogether in its construction. It is composed of three parts, all made of glass; the mixing bottle, Fig. 34; the Pipette, Fig. 35; and the burette, Fig 36. Besides, there should be ready three small glasses--one filled with tincture of litmus, the second with a solution of 1,369 per ammonia, and the third with the must or wine to be tested; also, a taller glass, or vessel, having its bottom covered with cotton, in which glass the burette, after it has been filled with the solution of ammonia, is to be placed in an upright position until wanted.

To use this instrument the must and the tincture of litmus, having first received the normal temperature of 14?Reaumer, are brought into the mixing bottle by means of the pipette, which is a hollow tube of glass, open on both ends. To fill it, place its lower end into the tincture or must, apply the mouth to the upper end, and by means of suction fill it with the tincture of litmus to above the line indicated at A. The opening of the top is then quickly closed with the thumb; by alternately raising the thumb, and pressing it down again, so much of the tincture is then allowed to flow back into the glass so as to lower the fluid to the line indicated at A. The remainder is then brought into the bottle, and the last drops forced out by blowing into the pipette.

"In filling it with must, raise the fluid in the same way, until it comes up to the line indicated at B, and then empty into the mixing bottle.

"The burette consists of two hollow tubes of glass. In filling it, hold the smaller tube with the right hand into the glass containing the solution of ammonia, apply the mouth to the larger one, and by drawing in the fluid the tube is filled exactly to the line indicated at 0 of the tube.

"Holding the mixing bottle by the neck between the thumb and forefinger of the left hand, place the smaller tube of the burette into the mouth of the mixing bottle, which must be constantly shaken; let enough of the solution of ammonia be brought drop by drop, into the mixture in the bottle, till the red has been changed into the deep reddish blue of the purple onion. This is the sign of the proper saturation of the acids. To distinguish still better, turn the mixing bottle upside down, by closing its mouth with the thumb, and examine the color of the fluid in the tube-shaped neck of the bottle, and afterwards, should it be required, add another drop of the ammonia. Repeat this until the proper tone of color has been reached, neither red nor blue. After thus fixing the precise point of the saturation of the acids, the burette is held upright, and the quantity of the solution of ammonia consumed is accurately determined,--that is, to what line on the scale the burette has been emptied. The quantity of the solution so used corresponds with the quantity of acids contained in the must--the larger division lines opposite the numbers indicating the thousandths part, and the smaller lines or dots the ten thousandths part.

"Until the eye has learned by practice to recognize the points of saturation by the tone of color, it can be proven by means of litmus paper. When the mixture in the bottle begins to turn blue, put in the end of a slip of litmus paper about half an inch deep, and then draw this end through your fingers, moistened with water. So long as the ends of the blue litmus paper become more or less reddened, the acids have not been completely saturated. Only when it remains blue, has the point of saturation been reached.

"In examining red must, the method should be modified as follows:--Instead of first filling the pipette with tincture of litmus, fill it with water to the line A, and transfer it into the bottle. After the quantity of must has been added, drop six-thousandths of the solution of ammonia into the mixture, constantly shaking it while dropping, then test it, and so on, until, after every further addition required with litmus paper, it is no longer reddened after having been wiped off."

DR. GALL further gives the following directions, as a guide, to distinguish and determine the proportion of acids which a must should contain, to be still agreeable to the palate, and good:

"Chemists distinguish the acid contained in the grape as the vinous, malic, grape, citric, tannic, gelatinous and para-citric acids. Whether all these are contained in the must, or which of them, is of small moment for us to know. For the practical wine-maker, it is sufficient to know, with full certainty, that, as the grape ripens, while the proportion of sugar increases, the quantity of acids continually diminishes; and hence, by leaving the grapes on the vines as long as possible, we have a double means of improving their products--the must or wine.

"All wines, without exception, to be of good and of agreeable taste, must contain from 4-1/2 to 7 thousandths parts of free acids, and each must containing more than seven thousandths parts of free acids may be considered as having too little water and sugar in proportion to its quantity of acids.

"In all wine-growing countries of Germany, for a number of years past, experience has proved that a corresponding addition of sugar and water is the means of converting the sourest must, not only into a good drinkable wine, but also into as good a wine as can be produced in favorable years, except in that peculiar and delicate aroma found only in the must of well-ripened grapes, and which must and will always distinguish the wines made in the best seasons from those made in poor seasons.

"The saccharometer and acidimeter, properly used, will give us the exact knowledge of what the must contains, and what it lacks; and we have the means at hand, by adding water, to reduce the acids to their proper proportion; and by adding sugar, to increase the amount of sugar the must should contain; in other words, we can change the poor must of indifferent seasons into the normal must of the best seasons in everything, except its bouquet or aroma, thereby converting an unwholesome and disagreeable drink into an agreeable and healthy one."

THE CHANGE OF THE MUST, BY FERMENTATION, INTO WINE.

Let us glance for a few moments at this wonderful, simple, and yet so complicated process, to give a clearer insight into the functions which man has to perform to assist Nature, and have her work for him, to attain the desired end. I cannot put the matter in a better light for my readers than to quote again from DR. GALL. He says:--"To form a correct opinion of what may and can be done in the manufacture of wine, we must be thoroughly convinced that Nature, in her operations, has other objects in view than merely to serve man as his careful cook and butler. Had the highest object of the Creator, in the creation of the grape, been simply to combine in the juice of the fruit nothing but what is indispensable to the formation of that delicious beverage for the accommodation of man, it might have been still easier done for him by at once filling the berries with wine already made. But in the production of fruits, the first object of all is to provide for the propagation and preservation of the species. Each fruit contains the germ of a new plant, and a quantity of nutritious matter surrounding and developing that germ. The general belief is, that this nutritious matter, and even the peculiar combination in which it is found in the fruit, has been made directly for the immediate use of man. This, however, is a mistake. The nutritious matter of the grape, as in the apple, pear, or any similar product, is designed by Nature only to serve as the first nourishment of the future plant, the germ of which lies in it. There are thousands of fruits of no use whatever, and are even noxious to man, and there are thousands more which, before they can be used, must be divested of certain parts, necessary, indeed, to the nutrition of the future plant, but unfit, in its present state, for the use or nourishment of man. For instance, barley contains starch, mucilaginous sugar, gum, adhesive matter, vegetable albumen, phosphate of lime, oil, fibre and water. All these are necessary to the formation of roots, stalks, leaves, flowers and the new grain; but for the manufacture of beer, the brewer needs only the first three substances. The same rule applies to the grape.

"In this use of the grape, all depends upon the judgment of man to select such of its parts as he wishes, and by his skill he adapts and applies them in the best manner for his purposes. In eating the grapes, he throws away the skins and seeds; for raisins, he evaporates the water, retaining only the solid parts, from which, when he uses them, he rejects their seeds. If he manufactures must, he lets the skins remain. In making wine, he sets free the carbonic acid contained in the must, and removes the lees, gum, tartar, and, in short, everything deposited during, and immediately after fermentation, as well as when it is put into casks and bottles. He not only removes from the wine its sediments, but watches the fermentation, and checks it as soon as its vinous fermentation is over, and the formation of vinegar about to begin. He refines his wine by an addition of foreign substances if necessary; he sulphurizes it; and, by one means or another, remedies its distempers.

"The manufacture of wine is thus a many-sided art; and he who does not understand it, or knows not how to guide and direct the powers of Nature to his own purposes, may as well give up all hopes of success in it."

So far DR. GALL; and to the intelligent and unbiased mind, the truth and force of these remarks will be apparent, without further extending or explaining them. How absurd, then, the blind ravings of those who talk about "natural" wines, and would condemn every addition of sugar and water to the must by man, when Nature has not fully done her part, as adulteration and fraud. Why, there is no such thing as a "natural wine;" for wine--good wine--is the product of art, and a manufacture from beginning to end. Would we not think that parent extremely cruel, as well as foolish, who would have her child without clothing, simply because Nature had allowed it to be born without it? Would not the child suffer and die, because its mother failed to aid Nature in her work, by clothing and feeding it when it is yet unable to feed and clothe itself? And yet, would not that wine-maker act equally foolish who has it within his power to remedy the deficiencies of Nature with such means as she herself supplies in good season, and which ought and would be in the must but for unfavorable circumstances, over which we have no control? Wine thus improved is just as pure as if the sugar and water had naturally been in the grapes in right proportions; just as beneficial to health; and only the fanatical "know-nothing" can call it adulterated. But the prejudices will disappear before the light of science and truth, however much ignorance may clamor against it. GALILEO, when forced to abjure publicly his great discovery of the motion of the earth around the sun as a heresy and lie, murmured between his teeth the celebrated words, "And yet it moves." It did move; and the theory is now an acknowledged truth, with which every schoolboy is familiar. Thus will it be with improved wine-making. It will yet be followed, generally and universally, as sure as the public will learn to distinguish between good and poor wine.

Let us now observe for a moment the change which fermentation makes in converting the must into wine. The nitrogeneous compounds--vegetable albumen, gluten--which are contained in the grape, and which are dissolved in the must as completely as the sugar, under certain circumstances turn into the fermenting principle, and so change the must into wine. This change is brought about by the fermenting substance coming into contact with the air, and receiving oxygen from it, in consequence of which it coagulates, and shows itself in the turbid state of must, or young wine. The coagulation of the lees takes place but gradually, and just in the degree the exhausted lees settle. The sugar gradually turns into alcohol. The acids partly remain as tartaric acid, are partly turned into ether, or settle with the lees, chrystallize, and adhere to the bottom of the casks. The etheric oil, or aroma, remains, and develops into bouquet; also the tannin, to a certain degree. The albumen and gluten principally settle, although a small portion of them remains in the wine. The coloring matter and extractive principle remain, but change somewhat by fermentation.

Thus it is the must containing a large amount of sugar needs a longer time to become clear than that containing but a small portion of it; therefore, many southern wines retain a certain amount of sugar undecomposed, and they are called sweet, or liqueur wines; whereas, wines in which the whole of the sugar has been decomposed are called sour or dry wines.

I have thought it necessary to be thus explicit to give my readers an insight into the general principles which should govern us in wine-making. I have quoted freely from the excellent work of DR. GALL. We will now see whether and how we can reduce it to practice. I will try and illustrate this by an example.

NORMAL MUST.

"Experiments continued for a number of years have proved that, in favorable seasons, grape juice contains, on the average, in 1,000 lbs.:

Sugar, 240 lbs. Acids, 6 " Water, 754 " ----- 1,000 "

This proportion would constitute what I call a normal must. But now we have an inferior season, and the must contains, instead of the above proportions, as follows:

Sugar, 150 lbs. Acids, 9 " Water, 841 " ----- 1,000 "

What must we do to bring such must to the condition of a normal must? This is the question thus arising. To solve it, we calculate thus: If, in six pounds of acids in a normal wine, 240 pounds of sugar appear, how much sugar is wanted for nine pounds of acids? Answer, 360 pounds. Our next question is: If, in six pounds of acids in a normal must, 754 pounds of water appear, how much water is required for nine pounds of acids? Answer, 1,131 pounds. As, therefore, the must which we intend to improve by neutralizing its acids, should contain 360 pounds of sugar, nine pounds of acids, and 1,131 pounds of water, but contains already 150 pounds of sugar, 9 pounds of acids, and 841 pounds of water, there remain to be added, 210 pounds of sugar, no acids, and 290 pounds of water.

By ameliorating a quantity of 1,000 pounds must by 210 pounds sugar, and 290 pounds water, we obtain 1,500 pounds of must, consisting of the same properties as the normal must, which makes a first-class wine."

This is wine-making, according to GALL'S method, in Europe. Now, let us see what we can do with it on American soil, and with American grapes.

THE MUST OF AMERICAN GRAPES.

If we examine the must of most of our American wine grapes closely, we find that they not only contain an excess of acids in inferior seasons, but also a superabundance of flavor or aroma, and of tannin and coloring matter. Especially of flavor, there is such an abundance that, were the quantity doubled by addition of sugar and water, there would still be an abundance; and with some varieties, such as the Concord, if fermented on the husks, it is so strong as to be disagreeable. We must, therefore, not only ameliorate the acid, but also the flavor and the astringency, of which the tannin is the principal cause. Therefore it is, that to us the knowledge of how to properly gallize our wines is still more important than to the European vintner, and the results which we can realize are yet more important. By a proper management, we can change must, which would otherwise make a disagreeable wine, into one in which everything is in its proper proportion, and which will delight the consumer, to whose fastidious taste if would otherwise have been repugnant. True, we have here a more congenial climate, and the grapes will generally ripen better, so that we can in most seasons produce a drinkable wine. But if we can increase the quantity, and at the same time improve the quality, there is certainly an inducement, which the practical business sense of our people will not fail to appreciate and make use of.

There is, however, one difficulty in the way. I do not believe that the acidimeter can yet be obtained in the country, and we must import them direct from the manufacturers, DR. L. C. MARQUART, of Bonn, on the Rhine; or J. DIEHN, Frankfort-on-the-Main.

However, this difficulty will soon be overcome; and, indeed, although it is impossible to practice gallizing without a saccharometer, we may get at the surplus of acids with tolerable certainty by the results shown by the saccharometer. To illustrate this, I will give an example:

Last year was one of the most unfavorable seasons for the ripening of grapes we have ever had here, and especially the Catawba lost almost nine-tenths of its crop by mildew and rot; it also lost its leaves, and the result was, that the grapes did not ripen well. When gathering my grapes, upon weighing the must, I found that it ranged from 52?to 70? whereas, in good seasons, Catawba must weighs from 80?to 95? I now calculated thus: if normal must of Catawba should weigh at least 80? and the must I have to deal with this season will weigh on an average only 60? I must add to this must about 1/2 lb. of sugar to bring it up to 80? But now I had the surplus acid to neutralize yet. To do this, I calculated thus: If, even in a normal Catawba must, or a must of the best seasons, there is yet an excess of acid, I can safely count on there being at least one-third too much acid in a must that weighs but 60? I, therefore, added to every 100 gallons of must 40 gallons of soft water, in which I had first dissolved 80 lbs. of crushed sugar, which brought the water, when weighed after dissolving the sugar in it, up to 80? Now, I had yet to add 50 lbs., or half a pound to each gallon of the original must, to bring this up to 80? I thus pressed, instead of 100 gallons, 150 gallons, from the same quantity of grapes; and the result was a wine, which every one who has tasted it has declared to be excellent Catawba. It has a brilliant pale yellow color, was perfectly clear 1st of January, and sold by me to the first one to whom I offered it, at a price which I have seldom realized for Catawba wine made in the best seasons, without addition of sugar or water. True, it has not as strong an aroma as the Catawba of our best seasons, nor has it as much astringency; but this latter I consider an advantage, and it still has abundant aroma to give it character.

Another experiment I made with the Concord satisfied me, without question, that the must of this grape will always gain by an addition of water and sugar. I pressed several casks of the pure juice, which, as the Concord had held its leaves and ripened its fruit very well, contained sugar enough to make a fair wine, namely, 75? This I generally pressed the day after gathering, and put into separate casks. I then took some must of the same weight, but to which I had added, to every 100 gallons, 50 gallons of water, in which I had diluted sugar until the water weighed 75? or not quite two pounds of sugar to the gallon of water, pressed also after the expiration of the same time, and otherwise treated in the same manner. Both were treated exactly alike, racked at the same time; and the result is, that every one who tries the two wines, without knowing how they have been treated, prefers the gallized wine to the other--the pure juice of the grape. It is more delicate in flavor, has less acidity, and a more brilliant color than the first, the ungallized must. They are both excellent, but there is a difference in favor of the gallized wine.

DR. GALL recommends grape sugar as the best to be used for the purpose. This is made from potato starch; but it is hard to obtain here, and I have found crushed loaf sugar answer every purpose. I think this sugar has the advantage over grape sugar, that it dissolves more readily, and can even be dissolved in cold water, thus simplifying the process very much. It will take about two pounds to the gallon of water to bring this up to 80? which will make a wine of sufficient body. The average price of sugar was about 22 cents per pound, and the cost of thus producing an additional gallon of wine, counting in labor, interest on capital, etc., will be about 60 cents. When the wine can be sold at from $2 to $3 per gallon, the reader will easily perceive of what immense advantage this method is to the grape-grower, if he can thereby not only improve the quality, but also increase the quantity of the yield.

The efforts made by the Commissioner of Patents, and the contributors to the annual reports from the Patent Office, to diffuse a general knowledge of this process, can therefore not be commended too highly. It will help much to bring into general use, among all classes, good, pure, native wines; and as soon as ever the poorer classes can obtain cheap agreeable wines, the use of bad whiskey and brandy will be abandoned more and more, and this nation will become a more temperate people.

But this is only the first step. There is a way to still further increase the quantity. DR. GALL and others found, by analyzing the husks of the grape after the juice had been extracted by powerful presses, that they not only still contained a considerable amount of juice, but also a great amount of extracts, or wine-making principles, in many instances sufficient for three times the bulk of the juice already expressed. This fact suggested the question: As there are so many of these valuable properties left, and only sugar and water exhausted, why cannot these be substituted until the others are completely exhausted? It was found that the husks still contained sufficient of acids, tannin, aroma, coloring matter, and gluten. All that remained to be added was water and sugar. It was found that this could be easily done; and the results showed that wine made in this manner was equal, if not superior, to some of that made from the original juice, and was often, by the best judges, preferred to that made from the original must.

I have also practiced this method extensively the last season; and the result is, that I have fully doubled the amount of wine of the Norton's Virginia and Concord. I have thus made 2,500 gallons of Concord, where I had but 1,030 gallons of original must; and 2,600 gallons of Norton's Virginia, where I had but 1,300 gallons of must. The wines thus made were kept strictly separate from those made from the original juice, and the result is, that many of them are better, and none inferior, to the original must; and although I have kept a careful diary of wine-making, in which I have noted the process how each cask was made, period of fermentation on the husks, quantity of sugar used, etc., and have not hesitated to show this to every purchaser after he had tasted of the wine, they generally, and with very few exceptions, chose those which had either been gallized in part, or entirely.

[Illustration: FIG. 37. UNION VILLAGE.--Berries 1/3 diameter.]

My method in making such wines was very simple. I generally took the same quantity of water, the husks had given original must, or in other words, when I had pressed 100 gallons of juice, I took about 80 gallons of water. To make Concord wine, I added 1-3/4 lbs. of sugar to the gallon, as I calculated upon some sugar remaining in the husks, which were not pressed entirely dry. This increased the quantity, with the juice yet contained in the husks to 100 gallons, and brought the water to 70; calculating that from 5?to 10?still remained in the husks, it would give us a must of about 80? The grapes, as before remarked, had been gathered during the foregoing day, and were generally pressed in the morning. As soon as possible the husks were turned into the fermenting vat again, all pulled apart and broken, and the water added to them. As the fermentation had been very strong before, it immediately commenced again. I generally allowed them to ferment for twenty-four hours, and then pressed again, but pressed as dry as possible this time. The whole treatment of this must was precisely similar to that of the original.

In making Norton's Virginia, I would take, instead of 1-3/4 lbs., 2 lbs. of sugar to the gallon--as it is naturally a wine of greater body than the Concord--and I aimed to come as near to the natural must as possible. I generally fermented this somewhat longer, as a darker color was desired. The time of fermentation must vary, of course, with the state of the atmosphere; in cooler weather, both pressings should remain longer on the husks. The results, in both varieties were wines of excellent flavor, good body, a brilliant color, with enough of tannin or astringency, and sufficient acid--therefore, in every way satisfactory.

The experiments, however, were not confined to these alone, but extended over a number of varieties, with good results in every case. Of all varieties tried, however, I found that the Concord would bear the most of gallizing, without losing its own peculiar flavor; and I satisfied myself, that the quantity in this grape can safely be increased here, from 100 gallons of must to 250 gallons of wine, and the quality yet be better, than if the must had been left in its normal condition.

And it is here again where only experience can teach us how far we can go with a certain variety. It must be clear and apparent to any one who is ever so slightly acquainted with wine-making, how widely different the varieties are in their characteristics and ingredients. We may lay it down as a general rule, however, that our native grapes, with their strong and peculiar flavors, and their superabundance of tannin and coloring matter, will admit of much more gallizing, than the more delicately flavored European kinds.

I have thus tried only to give an outline of the necessary operations, as well as the principles lying at the foundation of them. I have also spoken only of facts as I have found them, as I am well aware that this is a field in which I have much to learn yet, and where it but poorly becomes me to act the part of teacher. Those desiring more detailed information, I would refer to the Patent Office Reports of 1859-60, where they will find valuable extracts from the works of DR. GALL; and also to the original works.

If we look at the probable effect these methods of improving wines are likely to have upon grape-culture, it is but natural that we should ask the question: Is there anything reprehensible in the practice--any reason why it should not become general? The answer to this is very simple. They contain nothing which the fermented grape juice, in its purest and most perfect state does not also contain. Therefore, they are as pure as any grape juice can be, with the consideration in their favor, that everything is in the right proportion. Therefore, if wine made from pure grape juice can be recommended for general use, surely, the gallized wines can also be recommended. DR. GALL has repeatedly offered to pay a fine for the benefit of the poor, if the most critical chemical analysis could detect anything in them, which was injurious to health, or which pure wines ought not to contain, and his opponents have always failed to show anything of the kind.

I know that some of my wine-making friends will blame me for thus "letting the cat out of the bag." They seem to think that it would be better to keep the knowledge we have gained, to ourselves, carefully even hiding the fact that any of our wines have been gallized. But it has always been a deep-seated conviction with me, that knowledge and truth, like God's sun should be the common property of all His children--and that it is the duty of every one not to "hide his light under a bushel," but seek to impart it to all, who could, perhaps, be benefitted by it. And why, in reality, should we seek to keep as a secret a practice which is perfectly right and justifiable? If there is a prejudice against it, (and we know there is), this is not the way to combat it. Only by meeting it openly, and showing the fallacy of it, can we hope to convince the public, that there is nothing wrong about it. Truth and justice need never fear the light--they can only gain additional force from it. I do not even attempt to sell a cask of gallized wine, before the purchaser is made fully acquainted with the fact, that it has been gallized.

It is a matter of course, that many, who go to work carelessly and slovenly, will fail to make good wine, in this or any other way. To make a good article, the nature of each variety and its peculiarities must be closely studied--we must have as ripe grapes as we can get, carefully gathered; and we need not think that water and sugar will accomplish everything. There is a limit to everything, and to gallizing as well as to anything else. As soon as we pass beyond that limit, an inferior product will be the result.

But let us glance a moment at the probable influence this discovery will have on American grape culture. It cannot be otherwise than in the highest degree beneficial; for when we simply look at grape-culture as it was ten years ago, with the simple product of the Catawba as its basis; a variety which would only yield an average of, say 200 gallons to the acre--often very inferior wine--and look at it to-day, with such varieties as the Concord, yielding an average of from 1,000 to 1,500 gallons to the acre, which we can yet easily double by gallizing, thus in reality yielding an average of 2,500 gallons to the acre of uniformly good wine; can we be surprised if everybody talks and thinks of raising grapes? Truly, the time is not far distant--of which we hardly dared to dream ten years ago--and which we then thought we would never live to see; when every American citizen can indulge in a daily glass of that glorious gift of God to man, pure, light wine; and the American nation shall become a really temperate people.

And there is room for all. Let every one further the cause of grape-culture. The laborer by producing the grapes and wine; the mechanic by inventions; the law-giver by making laws furthering its culture, and the consumption of it; and all by drinking wine, in wise moderation of course.

WINE MAKING MADE EASY.

Some of my readers may think I did not look much to this, which I told them was one of the objects of this little work. To vindicate it and myself I will here state, that our object should always be to attain the highest perfection in everything. But, while I am aware that I have generally given the outline of operations on a large scale, I have never for a moment lost sight of the interests of those, who, like myself, are compelled, by bitter necessity, to commence at the lowest round of the ladder. And how could I forget the bitter experience of my first years, when hindered by want of means; but also the feelings of sincere joy, of glad triumph, when I had surmounted one more obstacle, and saw the path open wider before me at every step; and I can, therefore, fully sympathize with the poor laborer, who has nothing but his industrious hands and honest will to commence with.

While, therefore, it is most advantageous to follow grape-growing and wine-making with all the conveniences of well prepared soil, substantial trellis, a commodious wine cellar and all its appurtenances; yet, it is also possible to do without most of these conveniencies in the beginning, and yet succeed. If the grape-grower has not capital to spare to buy wire, he can, if he has timber on his land, split laths and nail them to the posts instead of wire. He can layer his plants even the first summer, and thus raise a stock for further planting; or dispose of them, as already mentioned in the beginning of this work. Or he can lease a piece of land from some one who wishes to have a vineyard planted on it, and who will furnish the plants to him, besides the necessary capital for the first year or so. I have contracted with several men without means in this manner, furnished them a small house, the necessary plants, and paid them $150 the first two years, they giving me half the returns of the vineyards, in plants and grapes; and they have become wealthy by such means. One of my tenants has realized over $8,000 for his share the last season, and will very likely realize the same amount next season.

And if he cannot afford to build a large cellar in the beginning, he can also do with a small one, even the most common house cellar will do through the winter, if it is only kept free from frost. One of our most successful wine-growers here, commenced his operations with a simple hole in the ground, dug under his house, and his first wine press was merely a large beam, let into a tree, which acted as a lever upon the grapes, with a press-bed, also of his own making. A few weeks ago the same man sold his last year's crop of wine for over $9,000 in cash, and has raised some $2,000 worth more in vines, cuttings, etc. Of course, it is not advisable to keep the wine over summer in an indifferent cellar, but during fermentation and the greater part of winter, it will answer very well, and he can easily dispose of his wine, if good, as soon as clear. Or he can dispose of his grapes at a fair price, to one of his neighbors, or take them to market.

But there is another consideration, which I cannot urge too strongly upon my readers, and which will do much to make grape-growing and wine-making easy. It is the forming of grape colonies, of grape-growers' villages. The advantages of such a colony will be easily seen. If each one has a small piece of suitable land, (and he does not need a large one to follow grape-growing), the neighbors can easily assist each other in ploughing and sub-soiling; they will be able to do with fewer work animals, as they can hitch together, and first prepare the soil for one and then for the other; the ravages of birds and insects will hardly be felt; they can join together, and build a large cellar in common, where each one can deliver and store his wine, and of which one perhaps better acquainted with the management of wine than the others, and whom all are willing to trust, can have the management. If there should be no such man among them, an experienced cooper can be hired by all, who can also manufacture the necessary casks. An association of that kind has also, generally, the preference in the market over a single individual, and they are able to obtain a higher price for their products, if they are of good quality.

There are thousands upon thousands of acres of the best grape lands yet to be had in the West, especially in Missouri, at a merely nominal price, which would be well adapted for settlements of that kind; where the virgin soil yet waits only the bidding of intelligent labor--of enterprising and industrious men--to bring forth the richest fruits. There is room for all--may it soon be filled with willing hearts to undertake the task.

And how much easier for you to-day, men with the active hand and intelligent brain, to commence--with the certainty of success before you--with varieties which will yield a large and sure return every year; with the market open before you, and the experience of those who have commenced, to guide you; with the reputation of American wines established; with double the price per gallon--and ten times the yield--compared with the beginner of only ten years ago, with nothing but uncertainty; uncertainty of yield, uncertainty of quality, of price, and of effecting a sale.

It took a brave heart then, and an iron will; the determination to succeed,--succeed against all obstacles. And yet, hundreds have commenced thus, and have succeeded. Can you hesitate, when the future is all bright before you, and the thousand and one obstacles have been overcome? If you do, you are not fit to be a grape-grower. Go toil and drudge for so many cents per day, in some factory, and end life as you have begun it. God's free air, the cultivation of one of His noblest gifts, destined to "make glad the heart in this rugged world of ours," is not for you. I may pity you, but I cannot sympathize with nor assist you, except by raising a cheap glass of wine to gladden even your cheerless lot.

###

Made in United States
North Haven, CT
23 November 2021

11450750R00054